T0258896

Population Dynamics of Commercial Fish in Inland Reservoirs

L.A. KUDERSKII

Translated from the Russian

A.A. BALKEMA/ROTTERDAM/BROOKFIELD/1996

Translation of:
Dinamika Stad Promyslovykh Ryb Vnutrennikh
Vodoemov, Nauka Publishers, Moscow, 1991

© 1996 Copyright Reserved

General Editor: Dr. V.S. Kothekar

ISBN 90 5410 259 4

Distributed in the USA and Canada by:
A.A. Balkema Publishers, Old Post Road, Brookfield, VT 05036, USA.

UDC 639:597 - 152.6 (28)

For the first time in ichthyological literature, the author has analyzed the structure of fish populations in inland reservoirs based on absolute values of their mass. The structure of population and its changes have been examined in seven fish species from small lakes, two from large lakes, and seven from large reservoirs in plains. Special attention has been paid to the main indicator of the structure of population—correlation between the age of ichthyomass culmination and mass maturation.

Data have been presented on age-related changes of ichthyomass in many generations of 11 species of commercially important fish. It is shown that populations of the investigated species of fish attain their maximum ichthyomass before reaching mass maturation or both these events are concurrent. Three types of population structure have been identified and their characteristics presented.

The material presented in the book and conclusions derived are of great interest both for developments in population ecology of fish and the theory of optimal fishing. The book is intended for ichthyologists, hydrobiologists, and ecologists studying theoretical bases of fish ecology and tackling applied questions of regulated fishing and conservation of fish stocks.

FOREWORD

In recent times, commercial fish of inland reservoirs have received great attention of ichthyologists. Regular investigations are being conducted on the determination of stocks of commercially valuable fish in order to develop forecasts of their possible catches as well as recommend rational exploitation of this resource in practically every large reservoir, lake or river basin. Studies on small lakes have also been intensified in connection with the search for ways of developing commercial fishery in them. The whole range of small lakes in the northwestern European USSR, Karelia, Urals, Western Siberia, Belorussia and other regions has been studied in depth toward this goal. Thanks to such works, a vast body of information has accumulated on different types of reservoirs, concerning different aspects of the functioning of aquatic ecosystems, including ichthyocenoses which are an integral part of these systems. Comprehensive material has been collected on many populations of individual commercial fish species.

Ichthyological material collected thus far enables us to move from numerous specific descriptions to a generalized analysis of different problems of the statics and dynamics of fish population in reservoirs. A confirmation of the propriety and fruitfulness of such investigations could be found in the reviews that have appeared in recent years on small lakes and on dynamics of ichthyocenoses in large lakes [*Change of Structure* . . ., 1982; Zhakov, 1984; Kitaev, 1984]. These works highlight several general problems of the structure of fish populations and its characteristics in connection with the production phenomena in aquatic ecosystems and variations of ichthyomass under the impact of anthropogenic factors. As regards the large range of problems concerning the analysis of peculiarities of the population structure of commercial fish and its dynamics, these do not find adequate mention in special literature.

Such a state of affairs is not incidental. Until recent decades in ichthyological and fishery investigations the traditional approach was predominant. This approach is based on the data of the analysis of the so-called average samples. Such data offer a possibility to evaluate individual processes in fish populations in terms of relative indices (percentage proportions of age-, size-, mass groups of fish, sex ratio, percentage proportions of different generations, etc.) or characteristics of average specimens and individuals (mean length and mass according to age classes, mean individual fecundity, etc.). However, the usefulness of a relative approach to population studies is rather limited. This approach does not provide an adequate idea about such cardinal population-structure indices as the ichthyomass of different age groups, its year by year dynamics and changes in individual reservoirs (population), ichthyomass of generations and its age-dependent variation and so on.

In recent decades, at par with the traditional approach in ichthyological and fishery investigations, extensive use is being made of the methods of absolute* (complete) determination of the number of fish in inland reservoirs. Thanks to this, a possibility has emerged to describe, using quantitative indices (ichthyomass, number of individuals), such aspects of the structure of populations which earlier remained undisclosed.

Several methods of absolute determination of the number of individuals and ichthyomass of reservoir fish are known and used in practice [Monastyrskii, 1952; Nikol'skii, 1974]. Among them, mention may be made of even such exotic methods as the determination of number of fish using data of number of eggs laid by perch [Zakhov, 1962]. In the present work, I have used the material obtained by two methods. One of them is the estimate of the number of fish after the ichthyocidal treatment of reservoirs. This method gives a large volume of information on ichthyomass of fish in small lakes. The method has been developed by E.V. Burmakin (1960, 1961, and so on) and is extensively used in many works of G.P. Rudenko (1962, 1967, 1978, and so on). It enables the study, with the maximum accuracy, of the various phenomena operating at the level of fish populations, including their structure and dynamics. The other method used in reservoirs, large and small lakes, is the well known method of areas. It is used in water reservoirs in the form modified by I.I. Lapitskii (1962, 1967), and in lakes—as modified by V.V. Gulin [Gulin, 1968; Sechin, 1977]. This method is recommended

---

*   By "absolute" we mean complete data on the number of fish and ichthyomass in a reservoir as opposed to "relative", based on the analysis of average samples. The term "absolute" should not be taken to mean absolute accuracy since methods used for such determinations, like all other methods of obtaining initial ichthyological data, are not free of errors, which depend on several factors [Poddubnyi, Malinin and Tereshchenko, 1982; Sechin, 1986].

as a basic method for fish estimates during evaluation of their stocks in commercial reservoirs [Sechin, 1986].

Thanks to systematic collection of absolute data for many reservoirs and lakes, information is available not only on the single-time estimates of the number of fish and ichthyomass but also multiyear results. These results make it possible to consider year-by-year variations of the structure of fish populations.

In the present work I have presented an analysis of the structure of fish populations from small lakes of Leningrad and Pskov regions, Lake Il'men' and reservoirs of the Volga Cascade and Tsimlyansk. The typological differences of reservoirs predetemined the structure of the described material. To maintain the logical uniformity of analysis, one chapter (Chapter 1) is wholly devoted to the examination of the structure of fish populations in small lakes. In Chapter 2, I describe in detail similar material on large water bodies (reservoirs and Lake Il'men'). Incidentally in each large water body there is a single population of each of the examined species of commercial fish. In reality, the matter is more complex [Poddubnyi, 1971; Poddubnyi and Malinin, 1988]. Such an assumption is valid in relation to small lakes but for water reservoirs, in general, it is not entirely correct. However, the question of finding local populations in individual water reservoirs, the data for which is used in the present work, remains as yet poorly studied.

In both chapters, to the best of my knowledge, such extensive information (based on absolute count) about the structure and dynamics of populations has been presented for the first time for many species of fish from different types of inland water bodies, both with developed and unexplored fishing. In view of this, the present work cannot be considered a monographic treatment of the problem of the dynamics of stocks of freshwater fish in general. Our aim is to present an analysis of new materials on the above problem, demonstrate the urgency and scientific importance of such works and thereby stimulate investigations aimed at further collection of specific data on the structure of populations of different commercial fish, the data based on absolute determination of the number of fish and ichthyomass in reservoirs.

Because of the above cited peculiarities of the work I have limited the literature citations only to those publications which have a definite bearing on the problem under discussion.

For this reason, the overall list of literature cited became short. Those who deal with dynamics of stocks of commercial fish very well understand that with a little desire and effort this list can be enlarged manyfold by including the numerous reviews as well as special papers. However, in my opinion this is not the need in the present case.

While preparing this work I used my individual publications as well as ones in coauthorship. All of them are listed in the literature cited. Besides published data, I have used primary material on absolute number of fish and ichthyomass determined by the Volgograd, Saratov, Tatarsk, Upper Volga, and Pskov divisions and Gorki laboratory of the State Scientific Research Institute of Lake and River Fisheries (GosNIORKh). I express my gratitude for the help, supply of material and participation in joint publications to Yu. V. Aleksandrov, V.I. Bandura, V.V. Baranova, S.A. Vetkasov, K.S. Goncharenko, V.G. Dronov, D.V. Zalozhnykh, V.N. Kartsev, I.I. Lapitskii, T.K. Nebol'sina, V.Ya. Nikandrov, Yu.I. Nikanorov, L.G. Perminov, A.S. Pechnikov, G.P. Rudenko, V.M. Tyunyakov, G.M. Fesenko, V.V. Khoruzhei, L.M. Khuzeeva, V.K. Chumakov, and S.V. Shibaev.

CONTENTS

CHAPTER 1

STRUCTURE AND DYNAMICS OF FISH POPULATIONS IN SMALL LAKES*

For various reasons, stocks of fish inhabiting small lakes are convenient for a study of their population structure. One reason is the diversity of types of these reservoirs, thanks to which the same species of fish is found under different habitat conditions. Particularly, the northwestern region, the material on whose lakes is examined here, represents a wide spectrum of fishery and limnological types of small lakes, like the ones examined in the recently published work of L.A. Zhakov (1984).

The second reason is the limited expanse of the water body. The area of these small lakes is tens or hundreds of hectares, rarely a few thousand hectares. Such are the sizes of lakes in which the population structure of fish was studied. Owing to the limited nature of the water area, one small lake inhabits a single species and, that too as its single population. Hence, as a result of investigation we obtain data in the most "pure" form. These data are free from irregularities which could arise due to heterogeneity of materials belonging to several populations of one and the same species; this cannot be ruled out while conducting work in large water bodies.

The third reason favoring investigations on small lakes concerning the population structure of fish is that many species, including the commercial ones, inhabit these water bodies [Berg, 1939; Lesnenko and Abrosov, 1973; Zakov, 1974; Gorbunova, Gulyaeva and Dmitrenko, 1978]. Furthermore, among small lakes there are water bodies which are inhabited by one fish species. This is usually the perch. A study of such fish populations, placed by nature in the unique conditions of existence,

---

* In this chapter I have used material contained in several papers, more importantly Kuderskii and Rudenko (1982, 1988), Kuderskii, Rudenko and Nikandrov (1983), Kuderskii, Aleksandrov and Perminov (1998), Kuderskii, Pechnikov and Rudenko (1988).

allows us to obtain rich information on the peculiarities of structure of populations, their dynamics and adaptational capacity.

Another equally important aspect of the population structure of fish from small lakes is their poor fishery exploitation. In most examples examined below, the commercial catch of fish was either absent or was of low intensity. In view of this, in many cases the examined fish populations were in an undisturbed state or closer to it at the time of these investigations.

Finally, the abundance of small lakes as opposed to large reservoirs make it possible to collect voluminous material. Accordingly, it is possible to conduct a broad comparative analysis and, in individual cases, even subject the data to statistical comparison.

Below I present the information on individual fish species. The populations of perch and bream have been examined in utmost detail, which is because of the nature of the obtained data. When for one and the same fish species the material was obtained after treating the lakes with ichthyocides and fishing with large fine-mesh catchnets, the same is described for each method of collection.

If data on the population structure of individual fish species were obtained from many lakes, the tables and illustrations give only some of them. The most typical examples were selected rather through subjective judgment. The reader may familiarize himself with the material on the remaining populations in my joint publications cited in the literature.

ROACH

The structure of roach populations was investigated in 13 lakes, of which six were treated with ichthyocides.

In small lakes under review, the roach populations are characterized by a long age series. The limiting age of fish observed in one of the lakes was 15+. However, often the age series terminates with individuals in the age 10+–12+ (9 out of 13 cases). Older age groups of roach are less numerous in all the populations and their ichthyomass is not large. The number of individuals and ichthyomass of age groups decrease sharply when fish reach the age of 4+–5+ (Table 1).

In all the small lakes investigated, the commercial catch of fish was either absent or was undertaken casually. Hence, the overall dynamics of the number of individuals and ichthyomass of age groups of fish is entirely controlled by natural causes as the groups become old. There may be three main causes: the conditions of multiplication, variable availability of food resources to individual age groups of fish, and intraspecific "predator-prey" type relationships. In small lakes the more

Table 1. Number of individuals and ichthyomass of age groups in roach populations from small lakes

Age	Lake Zhemchuzhnoe, 69 ha		Lake Somino, 21 ha		Lake Chernyavskoe, 65 ha		Lake Rubankovo, 82 ha		Lake Kudo, 161 ha		Lake Uzho, 120 ha	
	Number	kg	Number	kg	Number	kg	Number	kg	Number	kg	Number	kg
0+	284,100	994	689,900	1,290	1,548,886	2,788	694,123	347	5,512,423	11,025	660,228	792
1+	107,960	680	72,584	401	588,577	7,249	263,767	2,374	1,433,230	11,466	250,887	1,455
2+	48,580	505	6,626	155	264,860	4,750	118,695	1,424	429,969	6,020	112,899	937
3+	30,400	760	7,646	249	100,530	2,498	55,787	1,172	133,290	3,732	53,063	706
4+	18,310	601	2,170	98	9,694	793	24,546	884	42,653	2,175	23,348	546
5+	3,690	198	571	42	181	17	10,800	540	14,075	1,013	10,273	389
6+	1,020	80	117	14	108	13	3,996	252	4,786	450	3,801	207
7+	256	31	228	33	28	5	1,239	95	1,627	205	1,178	100
8+	128	23	2	0.5	20	5	347	29	553	92	330	41
9+	3	0.7	1	0.4	7	2	94	8	194	37	89	13
10+	2	0.6	1	0.4	3	1	24	3	74	19	21	4
11+	1	0.3	2	0.6	3	1	4	1	28	9	4	1
12+	—	—	1	0.6	1	0.4	—	—	12	4	—	—

Notes: 1. The first three lakes were treated with ichthyocides, in others fishing was done using large fine-mesh catchnet.
2. In this and later tables, under the column "number" single underline denotes age group attaining mass maturity; in the column under "kg" double underline denotes the culmination of ichthyomass in the age group of fish.

numerous perch and pike are the consumers of roach. These fish eat mostly the young age groups of roach.

In the investigated small lakes roach is numerous; this is primarily because of the favorable conditions for its multiplication. These lakes, as a rule, have an extensive shallow-water undergrowth zone which serves as a place for spawning of fish. Thus the overall structure of roach populations is formed under the impact of factors, on one hand, favoring appearance of more populous generations and, on the other, facilitating massive elimination, primarily of younger age groups of fish. Judged from the rates of decrease in numbers of fish with increase in age, the eliminating factors exert powerful influence on all populations of roach from small lakes.

It is interesting to note the variable rate of decrease in the number of individuals and ichthyomass with increase in age of fish. The rate of decrease in population seems to be higher. This can be seen from the value of the ratio of number of individuals and ichthyomass in the neighboring age groups of fish.

Below I illustrate this point from the example of roach population of Lake Rubankovo:

Age group of fish	0+,	1+,	2+,	3+,	4+,	5+,	6+,	7+,	8+,	9+,
	1+	2+	3+	4+	5+	6+	7+	8+	9+	10+
Ratio of number of individuals	2.6	2.2	2.1	2.3	2.3	2.7	3.2	3.6	3.7	3.9
Ratio of ichthyomass	0.2	1.7	1.2	1.3	1.6	2.1	2.7	3.3	3.6	2.7

This demonstration of the dynamics of number of individuals and ichthyomass of age groups with aging of fish is characteristic even for populations of roach from other small lakes. This can be readily verified from the data presented in Table 1.

In view of the variable rate of change due to the age of onset of mass maturation the number of roach decreases to 5/77–10/1867 whereas the ichthyomass decreases to 5/8–5/128* (in 13 populations). In the above difference of the rate of age-related changes, there is nothing unusual in both indices. As the age of fish increases, its number continuously and irreversibly decreases since replenishment occurs only from the zero age class. In dynamics of ichthyomass two diametrically opposite processes interact: continuous loss of ichthyomass under the impact of factors of natural mortality (and in populations caught, mortality during fishing in addition) as well as continuous increase of ichthyomass

---

* I have taken the number of age class 0+ and culminating ichthyomass of the age group as the starting point.

in connection with the individual weight gain continuing throughout the life of a fish.

The age of ichthyomass culmination is an important index characterizing the peculiarities of structure of fish population. In roach populations from small lakes, ichthyomass culmination is observed in the age classes 0+ and 1+ (Table 1). Such an early onset of culmination results, firstly, due to high level of reproduction of individuals and, secondly, by intensive elimination of fish in the young age. Here, the rate of elimination is such that it is not compensated for by the continuously occurring weight gain in fish.

The differences in the age of ichthyomass culmination in roach populations are not large (one year) and could be due to the dissimilar level of harvest of successive generations.*

Besides ichthyomass culmination the age of the onset of mass maturation** is an important criterion characterizing the specifics of the population structure of fish. In the investigated roach population, it was 3+ to 4+ or five full years [Table 1, Kuderskii and Rudenko, 1988].

In analysing the age-related dynamics of ichthyomass and considering the data on the age of the onset of mass maturation the following special features must be mentioned. In all roach populations, the maximum values of ichthyomass (culmination) of age groups come in the population of immature fish. Ichthyomass of the mature part of the roach population (according to the data of Table 1) is 2/5–5/148 of the ichthyomass of immature (and partly mature) age groups. Moreover, the ichthyomass of each mature age group is less. It is significant to note that the ichthyomass of any individually taken immature age group of roach is quite substantial. The concentration of populational ichthyomass in roach is completely or partly immature ages is clearly illustrated by the data presented in Fig. 1. All this once again emphasizes the high level of elimination observed in roach populations from small lakes. At the same time, such an intensity of elimination presupposes the presence of corresponding adaptations facilitating not only the survival but luxuriance of roach as a species. Hence, the relatively small ichthyomass of producers ensures stable production of reserves in roach populations.

Because of the concentration of ichthyomass in the sexually immature age groups its culmination in roach populations occurs 1⁻3 years before mass maturation of fish. In many lakes, ichthyomass culmination

---

* According to the published data [Pechnikov, Tereshenkov and Korolev, 1983; Pechnikov and Tereshenkov, 1984], in Lake Khvoinoe (area 100 ha) the age of ichthyomass culmination in roach populations was 1+, in Lake Naryadnoe (area 136 ha) it was 3+.

** In this work by mass maturation I imply such state of the age group of fish in which 2/3 and more females appear to be mature.

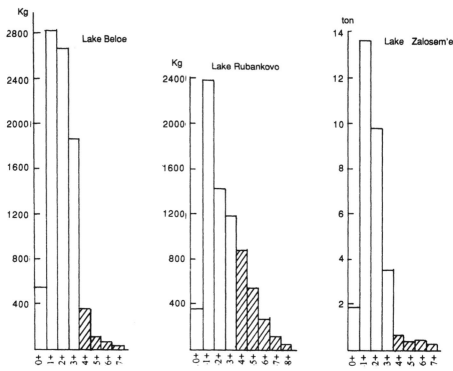

Fig. 1. Ichthyomass of age groups in roach populations from small lakes: Beloe (65 ha), Rubankovo (82 ha) and Zalosem'e (147 ha).
Age groups of fish starting from the age of mass maturation are identified with hachure.

in roach is observed even before the age of appearance of the first maturing females. Due attention has not been paid to this peculiarity of structure of roach populations. This has a great importance for understanding the course of production phenomena and cenotic relations in those communities of fish of which roach is a member.

From the totality of the data presented above and primarily from the correlation of the ages of ichthyomass culmination and mass maturation of females the roach populations from small lakes under review belong to the third type*. And roach could be considered as its typical representative.

From the data presented in Table 1 it may be seen that there is a similarity between the structure of roach populations from all the six lakes, although in three of them the material was obtained after treating the water body with ichthyocide and in the remaining three using large fine-mesh catchnets. From this a conclusion can be drawn that determi-

---

* The three types of population structures are fully described in Chapter 2 (see p. 71).

nation of the number of individuals of fish and ichthyomass by the area method is entirely reliable. The method gives completely representative idea about the structure of the investigated populations of fish.

BLEAK

Bleak is often found in small lakes but acquires commercial importance in a small number of water bodies. Data on the biology of bleak is very limited and there is hardly any information about the structure of its populations.

Because of the small commercial importance and low food quality bleak does not attract attention of investigators. However, when it is more numerous, its study is undoubtedly necessary for an objective assessment of the production processes in ichthyocenoses and lake eco-systems as a whole. The structure of bleak populations was studied in five small lakes. The data on the number of individuals of bleak and its ichthyomass were obtained by the area method.

The age series in the investigated bleak populations is rather re-stricted. The maximum recorded age was 6+–7+. Since during fishing in lakes with fine-mesh catchnets the older age groups of fish including those with short lengths are accounted for more completely than the younger groups, it may be considered that the mentioned age is close to the limiting maximum.

The number of individuals and ichthyomass in populations of bleak decreases sharply beginning with the age classes 4+–5+ (Table 2). In

Table 2. Number of individuals and ichthyomass of age groups in bleak populations from small lakes

Age	Lake Rubankovo, 82 ha		Lake Turichino, 109 ha		Lake Pelyuga (1975), 106 ha		Lake Pelyuga (1976), 106 ha	
	Number	kg	Number	kg	Number	kg	Number	kg
0+	117,982	236	295,195	590	314,300	629	138,200	64
1+	56,636	360	76,748	491	56,200	275	157,200	770
2+	31,150	436	23,024	299	106,600	1642	26,100	269
3+	16,821	320	7,138	121	42,800	740	53,100	988
4+	7,738	209	2,284	64	18,800	509	21,300	420
5+	2,786	103	754	28	3,600	131	9,400	351
6+	808	36	256	11	400	19	1,800	70
7+	—	—	87	4	—	—	—	—

Note: 1. The first three lakes were treated with ichthyocides, in others fishing was done using large fine-mesh catchnet.
2. In this and later tables, under the column "number" single underline denotes age group of fish attaining mass maturity; in the column under "kg" double underline denotes culmination of ichthyomass in the age group of fish.

other words, in this respect, bleak is similar to roach. However, if in roach for their low numbers and low ichthyomass of older age groups the series extends upto 12+ even to 15+, in bleak it already breaks after the 7+ group. Hence bleak may be considered, unlike roach, as a typical fish with short life cycle.

The rate of decrease of number of individuals in populations of bleak is higher than that of ichthyomass. This can be seen vividly from the following data (Lake Turichino):

Age groups of fish	0+,	1+,	2+,	3+,	4+,	5+,	6+,
	1+	2+	3+	4+	5+	6+	7+
Ratio of number of individuals	3.9	3.3	3.2	3.1	3.0	3.0	2.9
Ratio of ichthyomass	1.2	1.6	2.8	1.9	2.3	2.5	2.8

Such a regularity is observed also for populations of bleak from other small lakes. In view of this, the number of bleak by the age of the onset of mass maturation decreases to 10/31–5/646 while the ichthyomass decreases from the moment of culmination to 5/8–5/46 (from five populations).

Ichthyomass culmination in the investigated poulations of bleak is observed in the age classes 0+–3+, that is, it is more extended in comparison with roach. Of the five bleak populations, in two the age of ichthyomass culmination was equal to 3+ with overall age series extending to 6+. In roach, ichthyomass culmination was observed in the age 0+ and 1+ with duration of life cycle to 12+ and 15+. From these comparisons it can be concluded that the age of ichthyomass culmination is determined by ecological factors and is independent of the overall life span of the fish.

The differences in the age of ichthyomass culmination between bleak and roach are a consequence of their ecological specificity. As a planktophage, bleak has a more stable food base at all stages of its life cycle. Obviously, it is subject to pressure of predation to a lesser extent. Hence the age of ichthyomss culmination in bleak populations is, to a great extent, determined by the level of harvesting of individual generations. In particular, the high harvest (in comparison with the neighboring) generation of fish could remain in the stage of culminating group for several years. Moreover, the age of this culminating group shall keep on increasing.

In all the investigated bleak populations, mass maturation sets in at the age of four completed years. In view of this, ichthyomass culmination precedes mass maturation of fish by 1–3 years. Despite the fact that at times the mass maturation sets in a year after ichthyomass culmination, in all bleak populations the ichthyomass of immature age groups was 1.5–14.0 times as high as that of mature ones (Fig. 2). As shown by

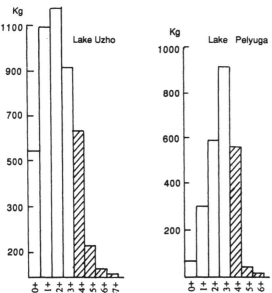

Fig. 2. Ichthyomass of age groups in bleak populations from small lakes Uzho (120 ha)
and Pelyuga (106 ha).
Age groups of fish starting from the age of mass maturation are identified with hachure.

Kuderskii and Dronov (1984) predominance of ichthyomass in the immature part of the population and more frequent onset of its culmination in age groups younger in relation to those having already attained mass maturation serves as the basis for including such populations in the third type. The investigated bleak populations, as a whole, satisfy these conditions.

WHITE BREAM

White bream is rare in small lakes. There are instances when alongside the resident form in small lakes, white bream which migrated from the neighboring large lake for reproduction is also caught [Kuderskii and Potapova, 1962]. However the local and migrant forms differ in their rate of growth and other features.

The structure of white bream population was studied in two lakes. The number of this fish contained in them was determined by the area method.

The maximum age of white bream in the investigated populations was 10 full years. However, this is not the limiting age for the bream in question. As in several other species of fish the rate of decrease of numbers in populations of white bream is higher than that of decrease of ichthyomass. This is clear from the data presented below (Lake Ostrovno-I):

Age group of fish	0+,	1,	2,	3,	4,	5,	6,	7,	8,	9,
	1	2	3	4	5	6	7	8	9	10
Ratio of number of individuals	2.6	2.2	2.1	1.0	3.1	3.3	2.9	2.9	2.3	1.5
Ratio of ichthyomass	0.3	1.0	1.5	0.7	1.8	2.2	2.2	1.8	1.8	1.2

Hence the number in age groups of white bream toward the onset of mass maturation decreases to 5/42–1/40 and the value of ichthyomass from culminating group by only 10/19–1/3.

Ichthyomass culmination in populations of white bream from the investigated lakes is observed in the age of two full years and the second at 4+ (Table 3). That means, the age of ichthyomass culmination in populations of white bream like bleak exerts influence on the level of productivity of individual (primarily intermediate) generations.

The age of onset of mass maturation in the investigated populations of white bream from small lakes is 5–5+. It is higher than the age of ichthyomass culmination. In view of this the ichthyomass of immature age groups in bleak population is 4.0–7.3 times as high as that of ma-

Table 3. Number of individuals and ichthyomass of age groups in white bream populations from small lakes

Age	Lake Ostrovno-I, 129 ha		Age	Lake Krivoe, 68 ha	
	Number	kg		Number	kg
0+	1,236,130	618	0+	50,814	56
1	469,730	1,879	1+	40,329	294
2	211,380	1,900	2+	32,007	397
3	99,349	1,291	3+	27,832	612
4	96,136	1,827	4+	24,204	777
5	31,059	994	5+	6,022	262
6	9,532	457	6+	334	18
7	3,287	204	7+	111	13
8	1,150	113	—	—	—
9	493	64	—	—	—
10	329	53	—	—	—

Note: 1.   The first three lakes were treated with ichthyocides, in others fishing was done using large fine-mesh catchnet.

2.   In this and later tables, under the column "number" single underline denotes age group of fish attaining mass maturity; in the column under "kg" double underline denotes culmination of ichthyomass in the age group of fish.

ture. Based on the estimates of the correlation of the age of ichthyomass culmination and mass maturation the white bream populations from the investigated small lakes could be included in the third type.

BREAM

In small lakes of the Northwest bream is usually a commonly found species of fish [Lesnenko and Abrosov, 1973; Shakov, 1984]. According to L.G. Perminov and Yu. V. Aleksandrov (1986), in the recent decades in Pskov Region the number of lakes inhabited by bream is continuously increasing. Unlike roach, bleak and white bream, which are small and so-called "slow growing" fish, bream is endowed with faster rate of growth and larger body size. In small lakes it is a valuable commercial fish, in view of which it was studied more intensively than the similar small ordinary fish species. Data on the structure of bream populations are available from 45 lakes. Estimate of the number of individuals and ichthyomass was done by the area method [Kuderskii, Aleksandrov and Perminov, 1988].

Bream belongs to species of fish with a long life cycle. The age series in populations of this fish from small lakes is quite prolonged and the fish attain the maximum age of 20+. In the predominant part of the population the maximum age is about 10+ and more.

Limiting age, years	7–7+, 8–8+	9–9+, 10–10+	11–11+, 12–12+	13–13+, 14–14+	15–15+, 16–16+	17–17+, 18–18+	19–19+ 20–20+
Number of populations	3	5	5	9	15	4	4

In 51.1% of bream populations, the investigated fish were found to be in the age of 15+–20+. Such a concentration of older age groups is observed not in every large water body.

Despite great stretch of age series the number of individuals and ichthyomass in older age groups decreases faster (Fig. 3). In many bream populations from small lakes, a sharp decrease of these indices is observed after the age of 5+–6+ but in some cases it occurs in much older fish (8+ and even 10+).

The intensity of utilization of bream reserves of small lakes is not high. In view of the organizational reasons and improper restrictions on catches of this fish, as envisaged by fishing laws ratified in 1961, a situation has developed in these lakes, which in essence tantamounts to neglect of fishery [Perminov, 1981; Perminov and Aleksandrov, 1986].

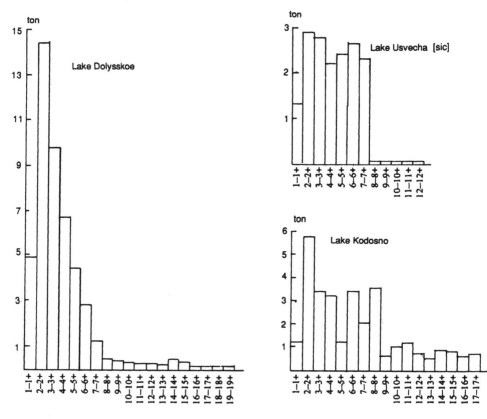

Fig. 3. Ichthyomass of age groups in bream populations from small lakes Dolysskoe (294 ha); Voronino-Lakhny (130 ha) and Kodosno (372 ha).

For this reason, the number and ichthyomass of bream attain higher values which can be seen from the data presented below*:

Number of individuals, thousands per ha	0.05–1	1–2	2–3	3–4	4–5	5–6	6–9	9–12
Number of populations	8	13	6	4	4	4	3	3
Ichthyomass, kg/ha	20–40	40–60	60–80	80–100	100–120	120–140	140–160	160–180
Number of populations	6	6	7	6	5	4	7	4

---

* The number of individuals and ichthyomass series for bream calculated per hectare of the reservoir were arrived at from the data of Perminov and Aleksandrov (1986).

On the average of 45 populations, the number of bream comes to 3027/ha and its ichthyomass 91.7 kg/ha [Perminov and Alksandrov, 1986] and is higher by one order of magnitude in respect of the number of fish in large reservoirs. The unusually high number of bream could not but affect the structure of its populations. It will be seen from the data presented in Table 4 that over the last 40 years the size and age structure of bream populations have undergone substantial changes. First, the age series has extended from 11+–12+ to 20+. Second, the body length and weight of similar age individuals has shown a marked decrease. Prolongation of the age series has led to complexity in the structure of bream populations on account of fish belonging to much older age classes, which were earlier not there.

The high number of bream in small lakes is due to two important factors; favorable conditions for multiplications and low rate of elimi-

Table 4. Size and age structure of bream populations from small lakes of Pskov Region [Source: Perminov and Aleksandrov, 1986]

Age	1948–1953		1969–1972		1974–1986	
	Length, cm	Weight, g	Length, cm	Weight, g	Length, cm	Weight, g
0+	5.6	2.9	—	—	5.4	4.1
1–1+	9.1	14.5	8.8	12.4	12.0	8.3
2–2+	13.6	57.6	10.4	28.1	11.9	47.5
3–3+	18.2	130.9	16.2	128.8	14.6	86.9
4–4+	23.1	266.0	18.6	193.4	16.9	119.0
5–5+	27.2	417.7	22.5	304.1	19.2	184.5
6–6+	30.8	620.7	24.1	350.6	21.9	264.0
7–7+	33.7	851.7	26.3	418.5	24.6	378.3
8–8+	36.7	1,084.5	29.9	622.2	26.6	461.7
9–9+	40.5	1,464.0	29.9	734.2	29.1	540.0
10–10+	40.9	1,692.6	37.9	1,251.6	30.1	648.3
11–11+	46.2	2,406.3	39.0	1,354.3	32.4	783.8
12–12+	46.0	2,536.5	—	—	34.2	919.0
13–13+	—	—	—	—	36.4	1,103.9
14–14+	55.0	3,970	—	—	38.5	1,274.9
15–15+	—	—	—	—	40.5	1,960.0
16–16+	—	—	—	—	41.5	1,541.8
17–17+	—	—	—	—	43.5	1,791.2
18–18+	—	—	—	—	45.7	2,108.3
19–19+	—	—	—	—	49.1	2,355.7
20–20+	—	—	—	—	44.6	2,000.0
Number of populations	28		13		50	

nation starting from the younger age groups. Hence, it may be said that in the small lakes investigated by us this species so to say luxuriates (in biological sense). However, its prospects in relation to the production [according to interpretation of P.A. Dryagin (1947)] seem highly restricted. Low elimination of younger age groups of bream is due to the relatively lower number of their consumers—pike. At the same time, for the perch which is numerous in small lakes, the young of bream seems unavailable primarily because of its greater body size.

In bream from small lakes, as in other species of fish, rate of decrease in the number of individuals in age groups with an increase in age is higher than the rate of decrease of the ichthyomass (Table 5). For this reason, by the time of mass maturation, the number of individuals in age groups decreases to 5/8–1/170. As regards the ichthyomass of age groups, unlike the earlier examined species of fish, in bream it changes differently. In the nature of change of ichthyomass, from its culmination to the age of the onset of mass maturation, all bream populations (out of 32 studied for the purpose) can be divided into three groups. The first group, the most numerous (14 populations), includes those populations in which the value of the ichthyomass by the age of mass maturation decreases to 10/11–5/26. In other words, for these bream populations the age related dynamics of the ichthyomass, which was mentioned for roach, bleak and white bream, is most typical.

At the same time, among bream populations another group (8 populations) is identified, in which the age of ichthyomass culmination coincides with the onset of mass maturation. In these populations, the ichthyomass culmination and ichthyomass of fish which reached mass maturity agree. Finally, there is the third group (10 populations), for which the increase of ichthyomass is typical even after the onset of mass maturation. In other words, the age of ichthyomass culmination is higher in them than in those which reached mass maturity. In populations belonging to this third group, ichthyomass from the age of mass maturation to its culmination increases 1.1–10.6 times and conversely the ichthyomass of fish which reached mass maturity is low by the same order (10/11–5/53). This aspect has been dealt in detail below.

The peculiarities of the size and age structure of bream populations in small lakes, observed in recent years (Table 4) in connection with a sharp increase in their numbers, cannot but affect such an index as the age of ichthyomass culmination and its correlation with the age of attaining mass maturity.

In bream populations from small lakes, ichthyomass culmination is observed in the age classs from 2–2+ to 11–11+. The data presented below gives an idea about the frequency of ichthyomass culmination according to individual age classes of fish:

Age of ichthyomass culmination	2–2+	3–3+	4–4+	5–5+	6–6+	7–7+	8–8+	9–9+ and older
Number of populations	10	3	6	12	8	3	—	3

Three maxima are observed in the number of populations depending on the age of fish. The first occurs in the younger age classes. Out of 45 populations, in 10 (22.2%) ichthyomass culmination occurs in the age class 2–2+. The second maximum includes the maximum number of populations and occurs in the age classes 4–4+ to 6–6+. In these age intervals ichthyomass culmination is observed in 26 populations (57.8%). Finally, the third maximum occurs in older ages (from 9–9+ to 11–11+) and covers 3 populations (6.7%).

Wide variation in the age of ichthyomass culmination in bream populations distinguishes this species from the ones discussed so far. This variation reflects greater structural diversity of bream populations inhabiting small lakes.

The age of ichthyomass culmination divides bream population into two parts. In groups of fish before culmination inclusively, there occurs an increase of ichthyomass of populations though this process is concurrent with the loss of a part of fish by natural (fishing) mortality. On the other hand, in much older age classes ichthyomass decreases despite

Table 5. Rates of decrease in number of individuals and ichthyomass of age groups in bream populations from small lakes

Constituent age classes	Lake Greater Vyaz, 276 ha		Lake Uzmen', 497 ha		Lake Sviblo, 1636 ha	
	Number	Ichthyomass	Number	Ichthyomass	Number	Ichthyomass
0+–1+	5.0	2.5	2.6	1.3	5.0	2.5
1+–2+	1.3	0.2	42.7	7.4	1.3	0.9
2+–3+	1.3	0.7	1.3	0.8	1.3	0.9
3+–4+	2.1	1.6	1.5	0.9	1.3	0.2
4+–5+	0.6	0.4	0.8	0.6	0.4	0.7
5+–6+	1.9	1.4	7.6	5.7	1.0	0.9
6+–7+	2.4	1.6	1.5	1.0	1.6	1.2
7+–8+	3.4	2.8	1.4	1.2	1.6	1.0
8+–9+	1.2	1.0	1.2	1.0	9.3	6.8
9+–10+	1.7	1.4	1.8	1.5	—	—
10+–11+	1.4	1.1	5.1	4.2	—	—

continuing weight increase of fish. Hence the change of position of culmination group according to the value of ichthyomass in the age series shifts the production process occurring in the populations as a whole. Because of this, two populations differing for example in the age of ichthyomass culmination strictly speaking, appear at variance in terms of production.

This last phenomenon is very clearly seen in bream populations from small lakes. Their diversity of population structure influences the ratio of ichthyomass of the immature and mature age groups. As mentioned above, in roach, bleak and white bream, the predominant part of the ichthyomass is concentrated in the immature (and partly mature) age classes. As opposed to these species, a complex picture is observed in bream. In half of the investigated populations (16), the ichthyomass of immature age groups was 1.05–4.62 times that of the mature. In other words, these populations of bream in the ratio of ichthyomass of mature and immature age groups behave same as those of roach, bleak and white bream. However, in addition to them, in 16 populations of bream from small lakes, the predominant part of the ichthyomass is found to be concentrated in the mature age groups. In such populations the ichthyomass of immature age groups is 0.04–0.82 of that of mature ones. In this manner the position of the ichthyomass culminating groups in the age series affects such productionally important indices as the ratio of total mass of immature and mature fish in a population.

Mass maturation of fish in populations of bream is observed in the age from 3–3+ to 8–8+:

Age of mass maturation	3–3+	4–4+	5–5+	6–6+	7–7+	8–8+
Number of populations	1	3	18	5	3	2

In this index of structure the investigated populations of bream show a distribution with the maximum in the middle age classes.

In bream populations the individuals maturing first are in the age from 2–2+ to 8–8+:

Age of first maturing fish	2–2+	3–3+	4–4+	5–5+	6–6+	7–7+	8–8+
Number of populations	3	10	13	3	2	—	1

In a majority of populations (23 or 71.9% of all investigated) the fish maturing first are found to be in the age classes 3–3+ and 4–4+.

Great stretch of the age of the onset of mass maturation at a cursory glance, may be correlated with the unusual overpopulation of bream in small lakes. However, a comparison of the age of attaining mass maturity of fish from entire population expressed in ichthyomass of bream in kg/ha, does not confirm such a conclusion (Table 6). In the dynamics of population density from 40 to 180 kg/ha, the investigated 32 populations of bream are distributed uniformly: 4–5 populations each in every density class. There is no zone of clustering of number of populations according to the age of the onset of mass maturation. Only in a comparison of the extreme age classes (3–3+ and 4–4+ on one side and 8–8+ on the other) it is possible to trace an increase in the age of maturity of fish depending on the ichthyomass of population. In bream populations with the age of attaining maturity starting at 3–3+ and 4–4+, the ichthyomass of 40–80 kg/ha (3 out of 4 populations) is predominant. When the age of attaining mass maturity is 8–8+, the ichthyomass reaches 120–180 kg/ha. However in the middle age groups (5–5+ and 6–6+) which account for 71.9% of the investigated bream populations, such a correlation is not noticed. The correlation of population density and the age of first maturing fish also behaves similarly.

Table 6. Distribution of bream populations from small lakes according to the density of population and the age of attaining mass maturity

Age of attaining mass maturity	Ichthyomass, kg/ha								n
	20–40	40–60	60–80	80–100	100–120	120–140	140–160	160–180	
3–3+	—	—	1	—	—	—	—	—	1
4–4+	—	2	—	—	1	—	—	—	3
5–5+	—	2	2	4	2	2	4	2	18
6–6+	1	—	1	—	—	1	1	1	5
7–7+	—	1	1	—	1	—	—	—	3
8–8+	—	—	—	—	—	1	—	1	2
Number of populations	1	5	5	4	4	4	5	4	32

Without going into the detailed analysis of this problem, which indeed requires special treatment, I would venture a very general conclusion that in the given case the age of onset of mass maturation in bream is not directly related to the density of its population but to some other ecological factors.

Data presented in Table 7 is of interest for understanding the mechanism of origin of structural peculiarities in different bream populations. In this table the distribution of bream populations is shown not in terms of the overall value of the ichthyomass but in terms of the percentage of ichthyomass shared by the culminating age group. It seems that in a vast majority of cases (26 populations or 57.8% of the total) the ichthyomass of culminating age group is less than 30% of the total ichthyomass of the bream population. In 9 cases (20%) it accounts for

Table 7. Distribution of bream populations from small lakes according to the proportion of ichthyomass of culminating age group and ichthyomass of bream population as a whole

Age of ichthyomass culmination	Ichthyomass of culminating age group as % of total ichthyomass of bream population										n
	0–10	10–20	20–30	30–40	40–50	50–60	60–70	70–80	80–90	90–100	
2–2+	—	4	2	3	—	1	—	—	—	—	10
3–3+	—	1	—	2	—	—	—	—	—	—	3
4–4+	—	3	1	1	—	—	—	1	—	—	6
5–5+	1	2	2	—	2	—	1	3	—	1	12
6–6+	—	3	2	1	—	1	—	—	—	1	8
7–7+	—	—	2	—	—	—	1	—	—	—	3
8–8+ and above	—	3	—	—	—	—	—	—	—	—	3
Number of populations	1	16	9	7	2	2	2	4	—	2	45

30–50% ichthyomass of the bream populations, in 8 (17.8%)—from 50 to 80% and only in 2 (4.4%) it is more than 90%. At the same time, no correlation is found between the relative ichthyomass of culminating fish group and its age in bream populations from small lakes.

In a study of the structure of fish population, of great interest is the analysis of the ratio of ages of culminating ichthyomass and mass maturation. All bream populations from small lakes, for which we know the age of mass maturation, can be divided into three groups. In populations of the first group the age of ichthyomass culmination precedes the age of mass maturation, in the second group they are concurrent and in the third the former succeeds the latter. In this regard, the bream populations differ greatly from the above examined populations of roach, bleak and white bream.

The distribution of the above identified three groups of populations according to the age of mass maturation of fish is given in Table 8. In the vast majority of populations (43.8% of 32 populations studied) ichthyomass culmination precedes the age of onset of mass maturation. If we take into account six more populations not included in Table 8 but for which the age of ichthyomass culmination appreciably precedes mass maturation, then the first group could include as many as 20 bream populations, which accounts for 52.6% of 38 populations. Thus for the entire totality of bream populations from small lakes the most characteristic group is the one in which the ichthyomass culmination occurs before mass maturation* of fish. It is this type of group that has been mentioned above for roach, bleak and white bream.

Table 8. Distribution of bream populations from small lakes based on the correlation of age of ichthyomass culmination and mass maturation

Age of mass maturation	Age of ichthyomass culmination			n
	Less	Equal	More	
3–3+	—	—	1	1
4–4+	1	1	1	3
5–5+	6	6	6	18
6–6+	4	1	—	5
7–7+	2	—	1	3
8–8+	1	—	1	2
Number of populations	14	8	10	32

---

* Of the 32 investigated bream populations, in 8 (21.9%) ichthyomass culmination preceded age classes in which first maturity of fish occurred.

The group of populations, in which the age of ichthyomass culmination and mass maturation coincides, is less numerous than the first (8 populations or 25.0%). In the absence of the third group and simplified approach to the analysis of the population structure it could be considered as a limiting case of populations of the first group. Such an assumption could even appear logical. However, actually the problem of correlation of the ages of ichthyomass culmination and mass maturation of fish is more complex. This is confirmed by the presence of a fairly large group of bream populations (10 or 31.3% of the total) in which ichthyomass culmination occurs after fish reach mass maturity.

In population of the first group ichthyomass culmination is observed sometimes directly before the year of onset of mass maturation, but often after 2, 3, 4 and 5 years. On the average of 14 populations the difference in the age of onset of both the phenomena was 2.4 years. In populations of the third group the culmination of ichthyomass is observed only in the year after the mass maturation of fish or succeeds it by 2, 3 or even 5 years. On the average of 10 populations this difference was found to be 2.0 years.

The structures of bream populations from small lakes, belonging to the third group are presented in Table 9.

It is of great interest to understand the explanation to the problem of relationships in bream populations in regard to the ichthyomass of immature and mature fish. As mentioned above, in population of roach, bleak and white bream a predominant part of the ichthyomass is concentrated in the immature (and partly mature) age groups. In bream populations a different, more complicated picture is observed. It seems that in populations in which the ichthyomass culmination occurs before mass maturation of fish a large part of the ichthyomass is shared by immature and partly mature* age groups. Out of 14 such populations in 12 we see such a correlation of ichthyomass of these two qualitatively different age groups. On the average of 14 populations the ichthyomass constituting age classes preceding mass maturation is 1.83 times as high as that of older age classes starting from the one in which mass maturation of fish is noticed. Predominance of ichthyomass of immature (and partly mature) fish in this group of bream populations can be clearly seen from the example of lakes Karatai, Nishcha and Soino (Fig. 4)

In populations of bream in which the ages of ichthyomass culmination and mass maturation coincide the correlation of ichthyomass of fish before and after mass maturation could be in favor of both immature (3

---

* Without going into details it must be mentioned that throughout in the text by the expression "immature (and partly mature) age groups of fish" we mean all age classes preceding the age of onset of mass maturation of fish.

Table 9. Number of individuals and ichthyomass of age groups in bream populations from small lakes

Age	Lake Loknovo, 636 ha		Lake Zhizhitskok, 5860 ha		Lake Orlovo, 800 ha		Lake Malkoe, 210 ha		Lake Nevel'skoe, 1400 ha		Lake Sviblo, 698 ha	
	Number in thousand	kg	Number in thousand	kg	Number in thousand	kg	Number in thousand	kg	Number in thousand	kg	Number in thousand	kg
1-1+	1,735.3	14,402	1,028.0	8,532	229.8	1,907	231.5	1,921	142.5	1,183	167.3	1,388
2-2+	1,193.6	56,696	370.5	19,821	136.6	6,488	102.7	4,878	115.9	2,898	131.5	2,235
3-3+	411.1	36,546	186.0	12,499	242.0	21,513	34.8	3,093	91.9	3,952	101.5	3,857
4-4+	561.4	66,862	111.1	10,132	370.4	44,114	23.6	2,810	70.5	4,535	80.1	4,325
5-5+	213.5	39,390	70.2	8,051	376.9	69,538	28.3	5,221	460.1	49,460	37.2	2,641
6-6+	61.4	15,104	43.5	6,250	86.1	21,180	7.7	1,894	344.1	59,701	18.6	1,711
7-7+	32.5	12,294	29.7	5,942	50.6	19,141	3.5	1,324	54.9	13,148	18.6	2,511
8-8+	8.6	3,970	19.8	5,805	28.7	13,250	3.1	1,431	15.5	5,750	27.2	5,630
9-9+	1.0	540	14.4	6,294	32.3	17,454	1.7	918	9.5	4,161	21.4	6,527
10-10+	0.8	518	10.4	6,739	4.9	3,189	2.2	1,426	6.0	3,600	24.3	9,558
11-11+	0.6	470	7.2	5,148	0.3	235	0.3	235	4.8	3,960	20.0	10,760
12-12+	0.3	275	5.7	4,337	0.3	248	0.2	183	4.8	5,160	12.0	7,356
13-13+	0.2	220	4.2	3,718	0.2	221	0.2	165	2.4	3,360	4.3	3,169
14-14+	0.1	127	3.2	3,347	0.1	255	0.04	50	1.2	1,920	1.4	1,286
15-15	0.1	196	2.5	3,070	0.04	78	0.01	19	1.2	1,920	0.6	676
16-16+	—	—	1.8	2,649	—	—	0.01	15	—	—	0.3	360

Note: In the column under 'kg' an underline denotes subculminating age group of fish in respect of ichthyomass. Other explanation same as in Table 1.

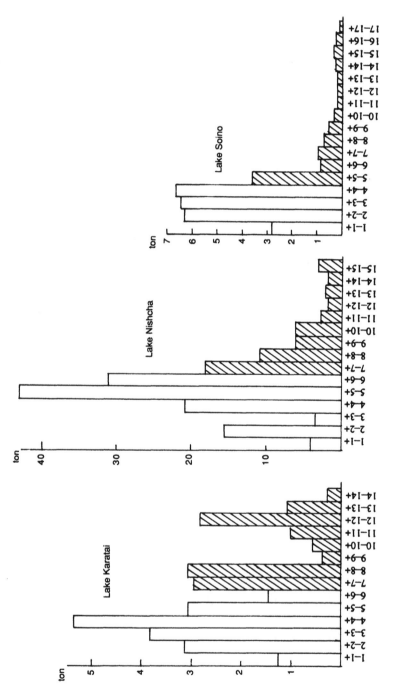

Fig. 4. Ichthyomass of age groups in bream populations from small lakes Karatai (549 ha), Nishcha (826 ha) and Soino (238 ha). Age groups of fish starting from the age of mass maturation are identified with hachure.

out of 8 populations) and mature age groups (5 out of 8 populations). On the average of 8 populations the ratio of ichthyomass of immature (and partly mature) age groups and mature groups works out to 0.83. In other words, a large part of ichthyomass of population is shared by mature fish. Thus, although in terms of the age of ichthyomass culmination and mass maturation, the fish populations of this group occupy a middle position among all the investigated bream population from small lakes, in the ratio of ichthyomass of immature (and partly mature) and mature fish, on the average, no equilibrium is observed.

In those bream populations in which ichthyomass culmination occurs after mass maturation of fish a large part of the ichthyomass of populations is, as a rule (in 9 out of 10 populations), concentrated in the mature groups, which can be very clearly seen from the example of Ushcho, Nechertsy and Sennitsa lakes (Fig. 5).

An analysis of the data presented in Table 9, Figs. 4 and 5 and corresponding section of the text makes it possible to conclude that in the correlation of age of ichthyomass culmination and mass maturation the bream populations from small lakes belong to the second and third types. Moreover, a transition from the third type which is more characteristic of bream to the second type occurs when generations arise, the level of productivity of which is substantially higher than those preceding them.

The transition of bream populations from the state corresponding to the third type to the second can be represented in the following manner. The harvest of the most productive generation appearing under favorable conditions occupies the position of the culminating fish group for a number of years and when a fairly large number passes through the age barrier of mass maturation to the region of the older age classes. In view of the continuously occurring processes of natural mortality such a generation ultimately loses the leading position and the next high-harvest generation becomes the culminating group. A period change occurs in the character of the structure of the population of bream and its transition again to the third type.

The multiyear dynamics of the structure of population of bream was not studied even in one small lake. However a large number of investigations of populations allows us to choose the most characteristic of them and place them in a sequential series which gives a vivid picture of how this process may occur. Figure 6 shows such a series for eight lakes. At the head of the series in Lake Uzho the ichthyomass culmination in bream populations occurs in the age classes 2–2+. In the succeeding lakes ichthyomass culmination shifts to older and older age classes and at the end of the series in Lake Usvai it occurs at the age of 10–10+.

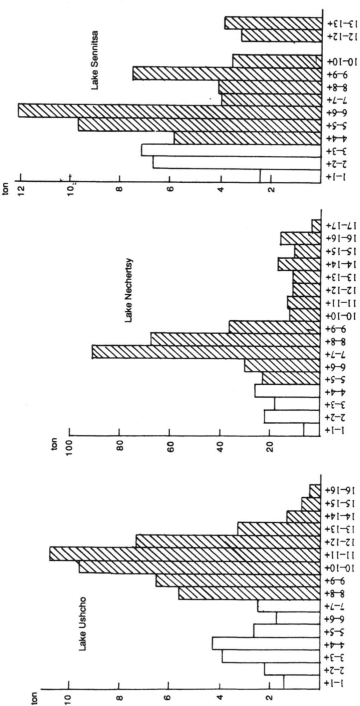

Fig. 5. Ichthyomass of age groups in bream populations from small lakes Ushcho (698 ha), Nechertsy (1969 ha) and Sennitsa (962 ha). Age groups of fish starting from the age of mass maturation are identified with hachure.

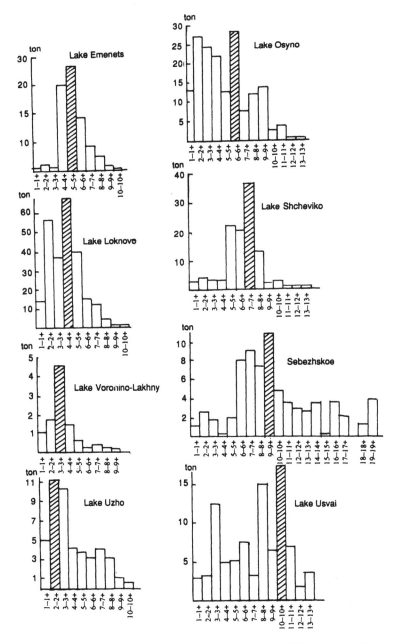

Fig. 6. Ichthyomass of age groups in bream populations from small lakes Uzho (805 ha), Voronino-Lakhny (130 ha), Loknovo (636 ha), Emenets (425 ha), Osyno (821 ha), Shcheviko (410 ha), Sebezhskoe (1599 ha), and Usvai (1954 ha). Age group with maximum (culminating) ichthyomass has been marked with hachure.

Then the cycle repeats itself if new high-harvest bream generations appear in the reservoir.

The above described scheme of periodic change of the structure of bream populations from small lakes may be accomplished at two levels: when the age of ichthyomass culmination precedes and succeeds the onset of mass maturation. The level of harvest of restructured bream populations at the first level is more common. It occurs under normal stress of biotic inter-relationships in the ecosystem and optimum rate of reproduction of numbers. With a change in biotic inter-relationships toward a sharp rise in the population (decrease in the pressure of predators, rise in the optimal rate of reproduction of numbers and so on) the periodicity of change of structure is attained at the second level, that is when the age of ichthyomass culmination is higher than that of attaining sexual maturity of fish. Such a change of structure is accompanied by a transition of populations from the third to the second type and vice versa.

From the point of view of production the restructuring of populations at the second level is irrational since the turnover of biomass in the ecosystem is significantly slowed down. Hence the production potential of the reservoir decreases.

The age-dependent changes in the value of ichthyomass in bream populations is quite variable in character. In some populations (for instance from Lake Loknovo—Table 9, Lake Olbito—Table 10) a gradual increase of ichthyomass is observed toward the culminating group and then there is an almost identical smooth decrease as the fish become old. Together with this there are populations in which besides the group with culminating ichthyomass there are subculminating older as well as younger groups of fish. Such groups could be one or many. For instance, in bream populations from lakes Zhizhitskoe, Nevel'skoe and Dniko there is one old subculminating group in each while in Lake Asho there are three. In bream population there is one young group in Lake Sviblo while in Lake Melkoe there are one young and one old subculminating groups (Tables 9 and 10).

The presence of subculminating groups according to the ichthyomass of age groups complicates the structure of bream populations. I have examined this problem in greater detail in the chapter on the structure of fish populations from large reservoirs in plains.

RUFF

Data on the structure of ruff populations was obtained for nine lakes including five after treatment with ichthyocides.

Table 10. Number of individuals and ichthyomass of age groups in bream populations from small lakes

Age	Lake Dniko, 515 ha		Lake Asho, 729 ha		Lake Olbito, 841 ha	
	Number in thousand	kg	Number in thousand	kg	Number in thousand	kg
1–1+	126.1	1,046	148.5	1,832	134.0	1,112
2–2+	98.8	4,693	115.5	3,811	116.0	5,510
3–3+	77.4	6,884	90.3	4,334	92.4	8,214
4–4+	62.5	7,443	70.3	4,639	195.0	35,977
5–5+	<u>113.7</u>	20,977	104.2	<u>13,858</u>	390.1	<u>95,964</u>
6–6+	164.0	<u>43,296</u>	<u>21.8</u>	3,422	273.0	67,158
7–7+	89.6	<u>33,918</u>	16.9	3,633	48.1	18,196
8–8+	10.0	4,617	28.9	<u>8,323</u>	16.9	7,802
9–9+	4.1	2,215	11.7	5,101	18.2	9,835
10–10+	1.8	1,166	5.4	2,511	9.1	5,899
11–11+	4.1	3,213	9.8	5,948	3.9	3,056
12–12+	8.8	8,087	9.1	6,206	2.6	2,389
13–13+	10.5	<u>11,590</u>	8.9	<u>7,476</u>	1.3	1,435
14–14+	4.1	5,227	5.0	4,815	—	—
15–15+	1.2	2,552	3.6	4,687	—	—
16–16+	1.7	2,621	1.2	1,563	—	—
17–17+	—	—	1.2	1,824	—	—
18–18+	—	—	0.9	1,350	—	—
19–19+	—	—	2.1	4,946	—	—

Note: See note to Table 9.

The age series in ruff populations is quite restricted. The maximum age of fish usually does not exceed 7+, in some lakes isolated individuals may be 8+. Still older individuals are not encountered.

In ruff populations the number in age groups decreases sharply after fish reach the age of 3+–4+ and even 2+ (Table 11). In their life span and rate of decrease of numbers and ichthyomass with age ruff populations are identical with those of bleak and are typical of species with a short life cycle.

Smaller lakes have been very little explored for fishing. As for ruff it is almost never caught. Hence the data presented in Table 11 and text refers, in essence, to nine populations. In this connection it is worth noting that the rate of decrease of fish population and ichthyomass with age is very high, the reasons for which are yet not clear. In my opinion, it is not correct to attribute the entire loss of numbers to predation in which perch predominate numerically.

Table 11. Number of individuals and ichthyomass of age groups in ruff populations from small lakes

Age	Lake Rubankovo, 82 ha		Lake Ozho, 120 ha		Lake Turichino, 109 ha		Lake Beloe, 65 ha		Lake Demenets, 66 ha		Lake Zhemchuzhnoe, 69 ha	
	Number	kg	Number	kg	Number	kg	Number	kg	Number	kg	Number	kg
0+	130,776	261	82,655	174	101,932	449	81,550	21	132,790	45	223,790	67
1+	37,925	303	32,235	103	38,734	271	23,850	48	9,044	18	64,900	136
2+	18,204	200	15,473	108	19,754	188	11,450	41	3,043	15	24,290	100
3+	10,012	170	8,510	81	10,509	150	4,400	34	584	6	4,990	31
4+	5,407	130	4,995	64	4,719	106	1,620	16	380	6	1,000	11
5+	2,487	67	2,114	47	1,293	40	130	1,5	186	3	3	—
6+	895	26	761	23	127	5	65	1,3	57	1,3	1	—
7+	260	8	296	12	—	—	1	—	5	—	1	—

Note: In the first three lakes fishing was done with large fine-mesh catchnet, the other three lakes were treated with ichthyocide. Legend same as in Table 1.

The rate of decrease of population and ichthyomass with age in ruff population is given below (for Lake Ruban-kovo):

Age groups of fish	0+, 1+	1+, 2+	2+, 3+	3+, 4+	4+, 5+	5+, 6+	6+, 7+
Ratio of numbers	3.5	2.1	1.8	1.9	2.2	2.8	3.4
Ratio of ichthyomass	0.9	1.5	1.2	1.3	1.9	2.6	3.3

Like other fishes, in ruff populations the rate of decrease of its numbers is higher than that of ichthyomass. In view of this by the age of onset of mass maturation the number decreases to 10/97–5/1137 while the ichthyomass decreases only to 5/7–2/15 (from 9 populations).

Ichthyomass culmination in ruff populations is observed in the youngest age classes: it was found in five cases in underyearlings (0+) and in four cases in two-year old (1+). In view of this, a major part of ichthyomass of these populations seems to be concentrated in the immature age groups. In nine populations the ichthyomass of immature (and partly mature) age groups was higher by 1.2–7.2 times and on the average by 3.3 times.

Ichthyomass culmination in ruff populations is observed 1–3 and even 4 years earlier than mass maturation of fish. Moreover it also precedes age classes in which maturation of fish occurs first. This fact must be emphasized because of the stocks of ruff, as mentioned above, are not exploited by commercial fishing in small lakes. Hence there is no need to consider the serious effect of fishing on the structure of ruff populations.

Considering the data presented above, ruff populations from small lakes could be included in the third group.

The data for the first three lakes presented in Table 11 were obtained using large fine-mesh catchnets for fishing, for the remaining three after the lakes were treated with ichthyocides. The structure of ruff populations in both cases shows monotypic character which confirms the reliability of the data obtained by the area method.

PIKE

Pike is found in most small lakes of the northwest but its number is usually not high. This is because pike as a predator consumes a large number of fish and is in need of game fishing in extensive individual areas. Emergence of overpopulation in pikes is practically ruled out in view of the cannibalism practiced by them. Pike inhabits lakes with multispecies ichthyofauna as also sometimes together with any one species of fish, more often perch. Below I examine pike populations only from lakes inhabited by several species of fish. I have used material from 14 small lakes, including six treated by ichthyocides.

The age series in pike populations is quite protracted. In some small lakes individuals up to 12+–14+ are found. However, the life span of pike is not the same in different reservoirs. Out of the 14 populations studied for the purpose only in three (21.4%) did it reach 10+–14+. In a vast majority of cases the age series in pike breaks at 6+–8+ (nine populations or 64.3%). The shortest age series in pike populations from small lakes terminates with age classes 4+ and 5+. The reasons for difference in the life span of pike in small lakes are varied and need special study with obligatory consideration of ecological peculiarities of lakes colonized by it.

The number of individuals and ichthyomass in various pike populations decrease dissimilarly with age. In some populations the decrease of these indices occurs gradually, in others their values decrease sharply (Table 12). Moreover this diversity of age-related dynamics of number of individuals and ichthyomass does not correlate with the overall life span of the fish.

In pike populations the rate of decrease of their numbers with age is higher than that of the ichthyomass (Table 13). A comparison of the data of Table 13 with similar data for earlier examined species of fish reveals the following special features: the rates of decrease of number of individuals and ichthyomass with age in pike is lower than in roach, bleak, white bream and ruff but is identical with that in bream. This is because both bream and pike in the investigated small lakes are, first of all, less prone to the effect of such factors as the "predator-prey" type relationships and secondly, they are endowed with high rates of individual weight gain.

In view of the variable rates of decrease in pike populations by the age of onset of mass maturation their number decreases to 10/31–5/521 while the ichthyomass (from ichthyomass culmination) to 0.9–0.33 (in 14 populations for numbers and 6 for ichthyomass)*. The rate of ichthyomass decrease in pike from the age of ichthyomass culmination to mass maturation is lower than in roach, bleak and ruff but is closer to similar index for white bream and those populations of bream in which ichthyomass culmination precedes mass maturation of fish. This difference is associated with long and short life cycles.

Such important index as the age of ichthyomass culmination, characterizing the structure of fish populations, varies in pike in wide range.

---

* From the calculation of decrease of the value of ichthyomass I have excluded Lake Uzho. Here, in the pike populations ichthyomass culmination occurs in fishes older than the age of mass maturation. I have also not considered populations with coinciding age of ichthyomass culmination and mass maturation.

Table 12. Number of individuals and ichthyomass of age groups of pike populations from small lakes

Age	Lake Uzho, 120 ha		Lake Kudo, 161 ha		Lake Krivoe, 68 ha		Lake Zhemchuzhnoe, 69 ha		Lake Somino, 21 ha		Lake Beloe, 65 ha	
	Number	kg	Number	kg	Number	kg	Number	kg	Number	kg	Number	kg
0+	7,406	170	846	23	1,243	34	1,175	6	5,732	15	3,430	10
1+	2,518	201	288	40	747	129	400	37	234	24	1,166	25
2+	1,335	191	152	46	449	141	190	42	300	85	618	64
3+	814	244	93	43	310	184	310	87	55	33	375	125
4+	529	264	78	70	243	214	70	51	42	15	150	99
5+	349	314	58	87	150	212	3	4	—	—	131	116
6+	213	298	36	69	43	88	1	3	14	31	38	44
7+	93	169	27	70	50	130	1	2	—	—	2	5
8+	—	—	9	27	62	170	—	—	—	—	—	—
9+	—	—	—	—	—	—	—	—	—	—	1	3
10+	—	—	—	—	18	67	—	—	—	—	1	5
11+	—	—	—	—	6	27	—	—	—	—	—	—
12+	—	—	—	—	6	23	—	—	—	—	—	—
13+	—	—	—	—	—	—	—	—	—	—	—	—
14+	—	—	—	—	6	66	—	—	—	—	—	—

Note : See note to Table 11.

Table 13. Rates of decrease of number of individuals and ichthyomass in pike populations from small lakes

Constituent age classes	Lake Uzho, 120 ha		Lake Kudo, 161 ha		Lake Turichino, 109 ha	
	Number	Ichthyomass	Number	Ichthyomass	Number	Ichthyomass
0+–1+	2.9	0.9	2.9	0.6	6.5	0.6
1+–2+	1.9	1.1	1.9	0.9	1.9	0.9
2+–3+	1.6	0.8	1.6	1.1	1.6	0.9
3+–4+	1.5	0.9	1.2	0.6	1.5	1.1
4+–5+	1.5	0.8	1.3	0.8	1.5	1.0
5+–6+	1.6	1.1	1.6	1.3	1.6	1.2
6+–7+	2.3	1.8	1.3	1.0	2.3	1.6
7+–8+	—	—	3.0	2.6	8.0	7.0

In the investigated populations it is in the range of 1+ to 5+*. In a vast majority of pike populations (12 out of 14, or 85.7%) ichthyomass culmination occurs in the age classes 3+–5+. However, in 2 cases it was found in groups of fish of younger age (1+ and 2+). Such a phenomenon is possible either with the appearance of highly productive generations or during intensive elimination (due to cannibalism) of fish in the age group of 2+, 3+ and older.

A considerable range of variation in the age of ichthyomass culmination confirms the great adaptability of pike to survival in the dynamic environmental conditions in small lakes.

The second important population-structure index is the age of onset of mass maturation in pike populations at 3+ to 5 complete years. In individual lakes the first maturing fish are in the age of 2+ but generally it is 3–4 years. The age of mass maturation of fish divides the population into two qualitatively differing parts. Younger fish constitute the immature and partly mature part of the population. Fish in the age of mass maturation and those older form the mature part of the population. In the roach, bleak and white bream and ruff examined earlier, the ichthyomass of immature (and partly mature) fish is higher than that of the mature ones. In pike populations in a majority of cases (10 out of 14, or 71.4%) this correlation is otherwise: the ichthyomass of age groups of fish preceding mass maturation is less than that of mature fish (from the age of mass maturation and older). The ratio of the first to the second ichthyomass is 0.11–0.93 and on the average 0.70. However, in four populations the ichthyomass of fish not reaching mass maturation is found to be 1.08–9.35 times as high as that of mature ones, average

---

* According to Pechnikov, Tereshenkov and Korolev (1983) in Lake Naryadnoe (area 136 ha) the ichthyomass culmination in pike populations occurs in the age class 0+.

being 3.61 times. Thus, while differing from roach, bleak, white bream and ruff in the ratio of the ichthyomass of immature and mature age groups of fish the pike populations are found to be similar in this index to bream. The difference lies only in that in bream the number of populations with predominance of ichthyomass of immature (and partly mature) fish correlates with the number of populations with predominance of ichthyomass of mature fish in the ratio of 1:1 while in pike it is 2:5. The difference in the ratios of both parts of populations is very high and explains the higher correlation of pike to the influence of such factors as fishing.

The relationship of age of ichthyomass culmination and mass maturation in pike are not monotypic. Out of the 14 populations in 7 (50.0%) ichthyomass culmination preceded mass maturation of fish, in 6 (42.9%) it occurred concurrently while in 1 (7.1%) it succeeded. In view of this, pike populations from some lakes belong to the third type while others are included in the second type. Transition is possible of one and the same population from the third type to the second and vice versa, depending on the harvest level of the individual generations and the degree of elimination of fish of younger age.

While studying the population structure of pike from small lakes the data on their number and ichthyomass were obtained by the area method and also by treating the lakes with ichthyocides. The data presented in Table 12 for the first three lakes were obtained by the area method and for the remaining three with the use of ichthyocides. It will be seen from these data that the area method gives satisfactory results in terms of the completeness of estimates of different age groups of pike.

PERCH

Perch is one of the most widely distributed and more populous species of fish in small lakes of the north-west. Together with roach it forms, in the reservoirs of this region, a unique perch-roach reserve. Thanks to their vast number, eurybiontic character and, in particular, euryphagy, perch often exercise decisive influence on the course of bioproductive processes in small lakes [Rudenko, 1967, 1971, 1978; Zhakov, 1968]. Their occurrence in numerous heterotypic lakes and predominance (or subdominance) in lacustrine ichthyocenoses results in a wide ecological plasticity of perch which is manifest also in the variability of structure of its populations depending on the conditions of existence [Kuderskii and Rudenko, 1982, 1988; Kuderskii, Rudenko and Nikandrov, 1983].

In perch the adaptive potential is exceptionally high and enables it to thrive even in those reservoirs in which it is the sole fish species.

Thereby perch is capable of inhabiting highly dystrophied lakes with exceptionally poor food base [Mel'yantsev, 1949; Kulikova, 1966]. It is possible to a priori assume that such extreme environmental conditions exert significant influence on the population structure of this species and its dynamics. Hence for a more objective analysis of peculiarities of the structure of pike populations all the small lakes inhabited by them should be divided into two groups. One of them includes lakes with multi-species composition of fish population while the other includes lakes inhabited by perch alone or by perch and pike. Inclusion of this biospecific combination with pure perch lakes is entirely justified by the aim of the present investigation.

Perch Populations from Lakes with Multi-species Fish Population

From the lakes with multispecies fish population I have studied 13 perch populations including ones from six reservoirs treated with ichthyocides. In all lakes perch is one of the numerous fish species.

The life span of perch in the investigated small lakes is 15+ but it differs from population to population. In more than half the populations (7 or 53.9%) the maximum age was 10+–12+. In some lakes (3 or 23.1%) fish older than 6–7 years were not found. But for all populations one characteristic feature was common; the contribution of older age groups was usually not high. A sharp decrease in their number and ichthyomass was observed, as a rule, in younger age classes (Fig. 7). In the degree of decrease of number of individuals and ichthyomass with age perch behaves similar to roach.

The similarity in the rates of elimination in the younger age groups between perch and roach owes its origin to the same factors: dissimilar availability of food to different age groups of fish and the pressure of predation. However, in perch populations these factors came to play somewhat earlier than in roach. In perch, in younger age, there is a transition from zooplankton feeding to feeding on nektobenthos and benthos and then to predation. However, in ultimate analysis, usually only a part of the population takes to predation. At the stages of changeover of the nature of feeding the pressure of trophic relationships intensifies, which is accompanied by higher elimination of the young ones. During transition to predation difficulties may arise for satisfying the food requirement, since the available young of other species of fish according to size is not always present among its sizes as determined by the degree of growth. In view of this, stressed trophic relationships arise, which lead to elimination of some fish. Moreover, among the young of perch even in the early age, individuals are identified accord-

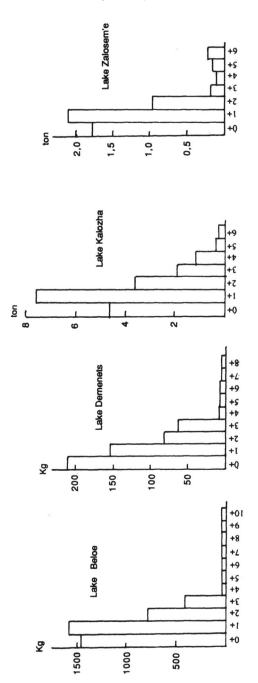

Fig. 7. Ichthyomass of age groups in perch populations from small lakes Beloe (65 ha), Demenets (6.6 ha), Kalozha (130 ha) and Zalosem'e (147 ha).

ing to the rate of growth, which very quickly change over to predation [Il'ina, 1970]. As already stated, stress in trophic relationships among young ones facilitates the appearance of such individuals. In predation the diet of perch includes not only the young of roach, bleak and other small fish but also the young of perch. Cannibalism increases the degree of elimination in the younger age groups of perch.

The consequence of the fact that unlike pike in perch only a part of the population takes to predation, is the lower number of individuals of older age groups with great prolongation of the age series. For example, in pike population from Lake Zhemchuzhnoe fish older than 5+ were found in small numbers and at the same time they attained the maximum age of 15+ and mass of up to 1 kg. A somewhat similar picture is observed in Chernyavskoe, Beloe and other lakes (Table 14, Fig. 7). As the biological analysis shows, in older age groups there is a predominance of fast growing perch, mostly predators. Thus perch populations often are ecologically heterogeneous. An analysis of their structure should, in essence, reveal the presence of two groups of fish; fast-growing with predatory feeding habit and relatively slow-growing with predominantly harmonious and mixed feeding habit. However, in the material at hand such a division was not attempted, in view of which the perch populations are considered below as a whole.

The rate of decrease of the number of individuals with age in perch populations as in other species of fish is higher than that of the ichthyomass (Table 15). According to the degree of decrease of these indices with age the perch populations differ from pike and similar populations of roach, bleak and ruff. That is to say, despite considerable prolongation of the age series, in such index as the degree of decrease of numbers and ichthyomass as also according to their sharp decrease in younger age, perch is closer to species of fish with a shorter life span.

In view of the differences in the rates of change of their numbers and ichthyomass by the age of mass maturation the former decreases to 2/41–10/3917 and the latter (according to the value of culminating ichthyomass) only 10/11–5/167 (from 13 populations).

From the data presented above it follows that the abundance of perch observed in small lakes is basically due to the large number of immature individuals. In populations of this species the mature individuals under the impact of processes of natural mortality are one–two orders of magnitude less. Despite small contribution, the mature part of the population successfully ensures replenishment of stocks of this fish.

In perch population from small lakes ichthyomass culmination is observed in the youngest age groups. Out of the 13 populations inves-

Table 14. Number of individuals and ichthyomass of age groups in perch populations of small lakes with multispecies fish population

Age	Lake Rubankovo, 82 ha		Lake Kudo, 161 ha		Lake Turichino, 109 ha		Lake Zhemchuzhnoe, 69 ha		Lake Somino, 21 ha		Lake Chernyavskoe, 65 ha	
	Number	kg	Number	kg	Number	kg	Number	kg	Number	kg	Number	kg
0+	1,140,667	912	257,015	771	363,937	1,201	174,540	384	218,840	482	111,540	547
1+	296,573	2,373	100,236	1,002	94,624	918	92,810	575	31,947	216	29,000	273
2+	88,972	934	45,106	902	28,387	329	20,370	440	9,802	128	15,630	242
3+	27,581	717	21,200	594	8,800	393	7,910	179	1,532	45	1,870	46
4+	8,826	282	9,328	951	2,816	296	4,220	140	239	14	440	17
5+	2,913	131	4,631	903	949	177	1,050	51	59	6	49	5
6+	990	59	1,912	621	323	97	37	3	77	21	18	7
7+	377	34	805	373	110	66	64	8	5	2	74	40
8+	114	15	367	213	37	27	37	6	1	0.8	95	64
9+	40	19	67	59	—	—	18	5	1	0.6	110	88
10+	15	10	16	16	—	—	8	2	—	—	37	33
11+	6	5	—	—	—	—	6	6	—	—	9	8
12+	—	—	—	—	—	—	13	3	—	—	18	23
13+	—	—	—	—	—	—	5	1	—	—	—	—
14+	—	—	—	—	—	—	1	1	—	—	—	—
15+	—	—	—	—	—	—	1	1	—	—	—	—

Note : See note to Table 11.

Table 15. Rate of decrease of number of individuals and ichthyomass of age groups in perch populations from small lakes

Compared age classes	Lake Uzho, 120 ha		Lake Ostrovno-I, 129 ha		Lake Demenets, 6.6 ha	
	Number	Ichthyomass	Number	Ichthyomass	Number	Ichthyomass
0+–1+	3.9	0.4	3.9	2.9	3.6	1.4
1+–2+	3.3	1.5	3.3	1.5	3.1	1.8
2+–3+	3.2	1.8	1.5	1.1	1.8	1.4
3+–4+	3.1	1.7	3.6	1.4	16.9	8.7
4+–5+	3.0	2.1	6.6	2.6	4.2	2.7
5+–6+	2.9	2.2	8.1	5.2	0.9	0.7
6+–7+	3.0	1.9	0.8	0.5	2.2	2.2
7+–8+	3.0	2.1	—	—	1.0	1.0
8+–9+	2.8	0.8	—	—	—	—
9+–10+	2.6	2.2	—	—	—	—

tigated, in 6 (42.2%) the fish were in the age group 0+ and in 7 (53.8%) in the age of 1+*. Such an early onset of ichthyomass culmination is necessarily considered as a direct consequence of the degree of elimination of younger age groups. In their age of ichthyomass culmination perch populations are similar to that of roach and ruff in which also it occurs in the youngest groups of fish.

The above discussion relating to the age of ichthyomass culmination refers to the slow growing individuals for which a homogeneous or mixed food is characteristic. As regards the fast-growing predator perch, obviously the periods of ichthyomass culmination are different. To elaborate on this question special investigations are necessary.

In comparing the age of ichthyomass culmination in different fish species the following conclusion becomes apparent. The population-structure index is determined by the specificity of the fish as well as to a significant extent by a combination of ecological factors of the niche. Among such factors as applied to perch of greater importance is the effectiveness of reproduction and the degree of elimination of the qualitatively different groups of individuals. In comparing a series of populations of one and the same species taken from different reservoirs the age of ichthyomass culmination may act as an indicator, making it possible to assess the suppression-favorable conditions or even blooming of fish in specific habitats.

---

* In Lake Naryadnoe (area 136 ha) according to the data presented in the work of Pechnikov, Tereshenkov and Korolev (1983) ichthyomass culmination in perch populations occurs in the age of 3+.

For perch populations it is characteristic that at ichthyomass culmination in younger age groups mass maturation sets in more older groups of fish. Of the 13 populations investigated in this respect, in 6 (46.2%) mass maturation was observed in fish at the age of 3+ and in 7 (53.8%) 4–5 completed years. Thus the gap between the ichthyomass culmination and mass maturation of fish is 2–4 years, in individual cases even 5 years. Throughout this period in perch populations both numbers and ichthyomass decrease continuously, although fish did not so far take part in reproduction (except some individuals from the age of 1 to maturity).

Thanks to the intensive elimination of fish in age classes preceding the onset of mass maturation in perch population, the ichthyomass of immature (and partly mature) fish was found to be considerably greater than that of mature fish. From a totality of all investigated populations the ichthyomass of immature fish was 1.9–35.8 times (on the average 10.4 times) as high as that of mature age classes. Such a significant predominance of immature individuals in the total ichthyomass is not found even in one of the investigated species of fish from small lakes. All this is intimately related with dual ecological nature of perch, which in the lacustrine ecosystems acts as a peaceful fish and as a predator and more so as a cannibal. It is possible to undoubtedly consider that in the ichthyofauna of reservoirs of the northwest there are other species which would be comparable to perch in all the above examined aspects. The great adaptive potential endows perch with a possibility not only to exist but form numerous populations even in the most inhospitable habitat conditions.

In all the investigated perch populations the age of ichthyomass culmination considerably preceded mass maturation of fish (Table 14). In many cases ichthyomass culmination was observed even before the age at which the first mature individuals appeared. Such a correlation of the age of ichthyomass culmination and mass maturation of fish is the direct consequence primarily of favorable conditions of reproduction and secondly of the intensive elimination of younger (mostly immature) age groups of perch.

In terms of all the above stated peculiarities the population structures of perch from small lakes belong to the third type. They are most similar (even in some details) to the populations of fish with short life cycle such as roach and ruff.

The material presented in Table 14 was obtained using the large fine-mesh catchnets for fishing in lakes (first three lakes) and with use of ichthyocide treatment. It is not difficult to see that both groups of populations have identical structures, which confirms the validity of the use of the area method to determine the numbers and ichthyomass of fish in reservoirs.

Perch Populations from Perch and Perch-pike Lakes

Thanks to euryphagy in a combination with cannibalism, perch is capable of inhabiting lakes in which, it is the only fish species and together with pike as well. Such lakes with extremely poor fish populations are often found in the northwestern region [Domrachev, 1922; Gerd, 1949; Zakov, 1974; Gorbunova, Gulyaeva and Dmitrenko, 1978]. Existence of perch in such conditions is possible in view of the fact that its young ones in the first year of their life feed on plankton, then changeover to feeding on benthic organisms. Perch begins predation very early [Popova, 1979], feeding in perch and perch-pile lakes on its own young members. Thus in the lacustrine ecosystems with extremely poor composition of fish, perch occupies ecological niches of both zooplanktophages and zoobenthophages as well as predators. Thereby it ensures transfer of energy from one trophic level to the other. Although in such conditions perch is the sole representative of fish, together with different age groups with dissimilar trophic status it is possible to consider perch population as a unique ichthyomass [Zhakov, 1984].

In the nature of overall influence on perch populations the perch pike lakes are quite close to pure perch lakes. In these lakes pike acts as a predator in relation to the perch population like the ichthyophagous perch*. Hence the peculiarities of structure of perch populations from such bispecific cenoses is considered in this section but after pure perch lakes.

Unique ecological conditions developing in pure perch lakes cannot but influence the structure of population of the species under reference. In order to elaborate on this problem I examined the material from five perch lakes treated with ichthyocides. All these lakes are very small in size, their reservoirs being 0.33–1.83 ha.

The life span of perch in these five investigated lakes is shorter than in reservoirs with multispecies ichthyofauna. It attains the maximum age of only 6+–9+. How characteristic is the shortening of life cycle in perch of all small lakes of similar type is difficult to judge from the data presented here. It is quite possible that the shortened age series is observed in some perch populations from perch lakes including the lakes investigated in this work. It is interesting here to note variable life span

---

*   In pike, cannibalism is very well developed and it affects the structure of its population roughly in the same manner as ichthyophagous perch affects the structure of perch population.

of perch of different populations. Since the reservoirs were treated with ichthyocides and the mortality of fish in this case was total, the possibility of noninclusion of older age groups is ruled out. Larger individuals, in this method, were accounted for totality. Hence the observed differences in the age structure of perch populations must necessarily be considered as a reliable fact. The differences are due to peculiarities of the processes of natural mortality in each specific reservoir since the effect of fishing is excluded because of its absence.

The numbers and ichthyomass of perch in the investigated five lakes are not high in terms of their absolute values (Table 16). However converted to per hectare of the reservoir the figures worked out to, numbers 767–2633/ha, weighted mean 1591/ha; ichthyomass 24.8–53.3 kg/ha, weighted mean 42.1 kg/ha. These figures are close to similar indices for small lakes with multispecies fish populations. In lakes mentioned in Table 14 the number of perch is 729–5200/ha, weighted mean 1901/ha; ichthyomass 13.0–55.8 kg/ha, weighted mean 30.0 kg/ha*. Thus, despite qualitative differences of the biotic processes occurring in lakes with multispeices ichthyofauna and in pure perch lakes such indices as the numbers and ichthyomass in both cases are identical when converted per unit area of the reservoir.

The rate of decrease of numbers in the perch population of pure perch lakes is higher than that of rate of decrease of ichthyomass. This is clear from the example of Lake Tyulen'e:

Age group of fish**	3+, 4+	4+, 5+	5+, 6+	6+, 7+	7+, 8+
Ratio of numbers	2.8	2.7	1.9	7.0	4.8
Ratio of ichthyomass	2.3	1.8	1.1	5.4	3.2

Similar are the rates of change of numbers and ichthyomass in perch populations from other small lakes (Table 16).

In all perch populations, data for which are presented in Table 16, the numbers in respective age groups change in a unique manner. Instead of the expected distribution with peak in the age group 1+ (underyearlings due to small sizes during collection of dead fish could be entirely excluded from purview) and following decrease in the older

---

* Since in perch lakes the age class 0+ is practically absent the data on lakes with multispecies fish population have been taken beginning with 1+ and older.
** Considering the peculiarities of the structure of perch populations from purely perch lakes the age groups with the maximum value of these indices have been taken as starting point for calculating the number and ichthyomass.

Table 16. Number of individuals and ichthyomass of age groups in perch populations from small perch lakes

Age	Lake Bezymyannoe (Leningrad Region), 120 ha		Lake Tyulen'e, 1.83 ha		Lake Chernyshok, 0.56 ha		Lake Bezymyannoe (Pskov Region), 0.33 ha		Lake Gulyak, 1.19 ha	
	Number	kg	Number	kg	Number	kg	Number	kg	Number	kg
0+	—	—	78	0.1	—	—	—	—	—	—
1+	110	0.5	264	3.0	43	0.9	20	0.6	171	3.5
2+	2,539	19.3	467	7.8	134	3.7	35	1.3	70	2.0
3+	275	6.8	1,494	36.5	203	6.9	154	7.0	50	2.1
4+	153	5.7	528	16.0	41	1.8	62	3.4	98	5.2
5+	56	2.8	194	9.1	8	0.4	60	3.7	213	13.7
6+	22	1.4	133	8.6	2	0.2	14	1.0	209	16.2
7+	4	0.3	19	1.6	—	—	-6	0.6	132	12.2
8+	—	—	4	0.5	—	—	—	—	65	6.7
9+	—	—	—	—	—	—	—	—	4	0.5

Note: Legend as in Table 1.

ages a different correlation is observed. Thus in Lake Bezymyannoe (Leningrad Region) the peak was observed in the age group 2+, in lakes Tyulen'e, Chernyshok and Bezymyannoe (Peskov Region) in 3+ and in Lake Gulshak 6+. Only in Lake Gulshak do we observe a second lesser peak for individuals in the age 1+. The observed peculiarity is not incidental and for populations from pure perch lakes it is due to wide trophic plasticity and cannibalism which is so characteristic of perch [Kuderskii, 1962; Gladkii, 1964; Balagurova, 1970; Popova, 1971, 1979].

E.V. Burmakin (1960) was the first to pay attention to this phenomenon. Based on the data on the estimate of perch in a small lake Plavushchee after the ichthyocide treatment, he wrote that the smaller number of small individuals of perch is rather confusing. In the subsequent work (Burmakin and Zhakov, 1961) this peculiar feature of the population structure of perch from pure perch lake was explained by him as due to low survival of young ones due to cannibalism. Burmakin's conclusion was confirmed by the data reported by G.P. Rudenko (1962).

In small lakes with multispecies fish population the structure of perch population follows the pattern common for fish species characterized by a sharp predominance of young ones in the age of 0+, 1+ (Rudenko, 1967, 1971, 1978; Kuderskii, and Rudenko, 1982). This is due to the fact that in lakes with multispecies ichthyocenosis different fish species play a unique buffer role in relation to each other and thereby weaken the pressure of predator (including perch) on each species individually. In perch lakes the pressure of predatory perch contrasts only its young ones.

V.V. Menshutkin and L.A. Zhakov (1964) while analysing the mathematical model of a monospecific perch population showed that its age structure changes periodically. Over several years a single generation occupies a dominant position, marked by a considerably larger number of young ones in it. As this generation starts becoming old the maximum number of individuals shifts to the side of older age groups. But simultaneously under the influence of natural mortality, the number of individuals decreases in such a generation and ultimately reaches such limits as would allow one of the new generations, despite losses due to cannibalism, to occupy a dominant position in the population. In this way there is periodic changeover of dominant generations and simultaneously change of structure of perch populations in lakes in which it is the only fish species (Fig. 8).

On the model examined by V.V. Menshutkin and L.A. Zhakov (1964) they established theoretically possible picture of change of the age structure of perch population. At that time these authors did not have at their disposal factual material. The data presented in Table 16 confirm the above theoretical model. All the five perch lakes naturally split into three groups for which the maximum number of individuals is found in

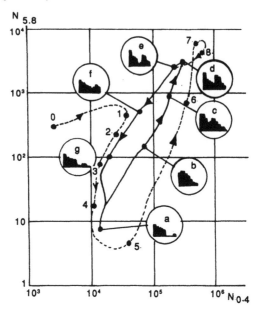

Fig. 8. Phase diagram of perch population.
On abscissa—number of immature individuals; on ordinate—number of mature individuals. Thick line denotes the established cycle, broken line denotes the transitional process; a–g—histograms of age composition of population [Source: Menshutkin, 1971].

the age of 2+, 3+ and 6+. In other words, each of these three groups of populations corresponds to different states of the model. This also confirms the basic principle accepted in the construction of this model which lies in recognizing the leading role on infraspecific interrelationships (cannibalism) in the population dynamics of perch from pure perch lakes [Kuderskii, Rudenko and Nikandrov, 1983].

In view of the above stated dynamics of structure, the reduction in the number of individuals and ichthyomass to the time of mass maturation of fish in such perch populations occurs differently than in lakes with multispecies fish population. All the five populations from perch lakes break into three groups. In one of them the number of individuals and ichthyomass decrease from the maximum values to the age of mass maturation of fish respectively to 5/14–5/83 and 10/23–5/27 (Lake Bezymyannoe, Leningrad Region and Lake Tyulen'). In the second group there is only Lake Chernyshok. In the perch populations living in it, the age of maximum value of the number of individuals and ichthyomass and mass maturation occur concurrently and consequently the ratio of these indices is 1.0. In the third group there are two populations of perch (Lake Bezymyannoe, Pskov Region and Lake Glushak) in which the maximum value of the number of individuals and of ichthyomass is observed after the age of mass maturation. Both indices from the moment of mass maturation to the age of attaining maximum values increase respectively 2.2–4.4 and 3.1–5.4 times. In the nature of change of the number of individuals and ichthyomass from maximum values by the age of onset of mass maturation the perch populations from pure perch lakes differ substantially from populations in lakes with multispecies ichthyofauna and are similar to populations of such species sharply differing from perch, like bream.

A confirmation of the above cited data about the heterogeneity of perch populations from perch lakes is offered by the ratio of ichthyomass of immature and mature age groups. In two populations the ichthyomass of immature (and partly mature) fish is higher than that of mature fish by 1.32–2.53 times. In three populations it is less, 0.96–0.5. In this ratio populations of perch lakes differ substantially from the ones from multispecies fish populations. In the latter, as already stated earlier, the ichthyomass of immature age groups is considerably higher than that of mature ones.

Ichthyomass culmination in the above perch populations (Table 16) occurs in the age classes 2+–3+ and even 6+. In this respect they differ considerably from perch populations from lakes with multispecies fish populations in which ichthyomass culmination is observed in age classes 0+–1+.

Peculiarities of habitat conditions in perch lakes influence (at least slightly) also the age of the onset of mass maturation in fish. Thus in Lake Bezymyannoe (Pskov Region) mass maturation of female perch is observed in the age 2+. However in the remaining four lakes mass maturation in perch populations is observed in the age of 3+–4+; that is practically in the same period as in perch from reservoirs with multispecies ichthyofauna.

In view of the change in the age of ichthyomass culmination in perch populations from perch lakes different correlation (in comparison with populations from lakes with multispecies fish population) of this index is established with the time of onset of mass maturation. If in lakes with multispecies ichthyofauna the ichthyomass culmination always sets in before mass maturation of fish and the time gap between these events is usually larger than in populations from perch lakes a different picture is observed. In these cases ichthyomass culmination either precedes, coincides or succeeds the mass maturation of fish. In other words it occurs in some populations in immature and in others in mature age groups (Table 16). Such is the influence of specific ecological conditions on such important population-structure index as the correlation of the age of ichthyomass culmination and mass maturation.

Thus perch population from small lakes inhabited by only this species could belong to the third type of structure as well as to the second type. Moreover the structure and its associated affinity to one or the other type changes cyclically. These changes are not random and owe their origin to such objectively operating factors as the extent of survival of small fish which depends on the degree of cannibalism being manifested. Under a relatively low numerical strength of large fish the "pressure" of cannibalism decreases and a large number of small individuals would be present in a population. A rise in the number of individuals and ichthyomass of larger fish, thanks to individuals that were earlier small sized, leads to an increase of the "pressure" of cannibalism and a corresponding decrease of the number of individuals and ichthyomass of younger age. The structure of perch populations from small lakes inhabited by only this species oscillates as if between two extremes; one of them may be called a rejuvenated state, and the other, aged. In the rejuvenated population, young individuals predominate, and ichthyomass culmination is observed before the moment of mass maturation (population of the third type). In the aged population, ichthyomass culmination occurs after attaining mass maturity (population of second type).

The analysis of these extreme states of perch populations is of definite theoretical interest from the production viewpoint. In such an approach a population may be divided into two groups: one of them (small individuals) feed on invertebrates and because of this, produce the perch biomass. The other (large individuals), thanks to the monospecific composition of ichthyocenose, feed not only on invertebrates but also on perch. The larger the fish the greater the degree of their cannibalism. Hence the second group while producing some quantity of perch ichthyomass simultaneously destroys a definite part of the perch population. In an extreme case the first group adds the population ichthyomass and the second spends it. At the same time the second group ensures reproduction of population as a whole including individuals of the first group. In the ultimate analysis there arises a unique, rarely encountered situation needing elaboration in the treatise on general ecology.

Perch population in its structural relationship was studied also in the five perch-pike lakes treated with ichthyocides. All of them are small lakes in the true sense as their area is 2.74–15.60 ha.

In the perch-pike lakes the perch populations are in some respect similar to populations from pure perch reservoirs but differ in many indices from the latter. Links between fishes in the perch-pikes lakes become complex. Perch populations are under the influence not only of individuals of their own kind, but also of such active predators as pike. On the other hand, pike exerts great influence on its own population since the large individuals of this species eat the smaller ones. Hence pike regulates its own number. On the whole, the influence of pike on perch populations is similar to the influence of perch on its own populations. But the presence of an additional predator endowed with several ecological specificities adds all the same, certain new features to the structure of perch populations.

The first thing that attracts our attention is the large variation in the range of the age series of perch in individual lakes (Table 17). If in Lake Svetloe it extends to 12+, in Lake Vyselki it terminates at the age of 5+. Moreover perch populations from Lake Vyselki and Kalozha include only five age groups of fish of which respectively one and two are represented by isolated individuals. In population from Lake Kalozha and Belevets the first two-three age groups are absent. In relation to underyearlings their incomplete estimate is still permissible because of small size. However dead fish in the age of 1+ and 2+ during chemical treatment of the lakes were counted completely. Hence their absence could be interpreted as disappearance of these generations, thanks to consumption by predators.

Despite intensive feeding on perch by predators its number of individuals and ichthyomass in the perch-pike lakes is not lower than in pure perch lakes and in lakes with multispecies fish populations. Thus in Lake Vyselki the number of perch was 3353/ha and ichthyomass 52.6 kg/ha; in Lake Krivenets 3748/ha and 57.9 kg/ha, in Lake Svetloe 1780/ ha and 51.9 kg/ha, in Lake Kalozha 959/ha and 26.5 kg/ha and only in Lake Belevets were these figures lower—42.2/ha and 14.5 kg/ha. The mean weighted values of these two indices for all the five lakes are 2102/ha and 46.6 kg/ha. From the data reported here it may be noted that despite high specificity of the ecosystem of perch-pike lakes and peculiarities of biotic processes taking place in them, the perch populations inhabiting them are not inferior, in terms of production, to other types of lakes. This latter fact is, as mentioned above, due to wide ecological plasticity of perch and its ability to establish higher number of individuals and ichthyomass even in extreme conditions of habitation.

The age of ichthyomass culmination in perch populations from perch-pike lakes changes from 2+ up to 4+ and even reaches 7+ (Table 17). In this respect the examined populations have been investigated in detail from the perch lakes and differ greatly from populations in lakes with multispecies ichthyofauna.

The change in the number of age groups of populations from perch-pike lakes in general is similar to that reported for perch reservoirs. However there are some unique features. In Lake Vyselki one observes a decrease in the number of individuals with age, common for most fish species. But in lakes Krivenets and Svetloe the nature of change in the number of individuals of different age groups is similar to that reported for perch lakes. Finally in lakes Kalozha and Belevets the younger age

Table 17. Number of individuals and ichthyomass of age groups in perch populations from small perch-pike lakes

Age	Lake Vyselki, 3.45 ha		Lake Krivenets, 7.70 ha		Lake Svetloe, 15.60 ha		Lake Kalozha, 2.74 ha		Lake Belevets, 4.27 ha	
	Number	kg	Number	kg	Number	kg	Number	kg	Number	kg
0+	—	—	1,905	9.5	139	0.5	—	—	—	—
1+	7,880	58.3	7,230	72.2	272	0.7	2	—	—	—
2+	3,515	111.8	9,279	130.5	7,719	89.5	—	—	—	—
3+	135	8.0	8,694	170.4	13,516	289.2	462	9.2	7	0.7
4+	31	2.5	1,030	32.0	3,870	149.8	2,526	58.9	11	1.6
5+	5	0.9	368	14.2	,520	26.8	93	4.1	17	2.9
6+	—	—	170	7.7	570	43.7	6	0.5	17	4.8
7+	—	—	138	7.9	830	79.0	—	—	58	20.6
8+	—	—	14	1.7	235	31.2	—	—	38	13.6
9+	—	—	—	—	182	43.9	—	—	32	17.8
10+	—	—	—	—	57	23.3	—	—	—	—
11+	—	—	—	—	42	21.0	—	—	—	—
12+	—	—	—	—	20	11.5	—	—	—	—

Note: Explanation same as in Table 1.

groups are absent in the perch populations. Despite these unique features, all populations of perch from perch-pike lakes in general follow the scheme of cyclic changes of structure described above for pure perch lakes. All the five perch-pike lakes form a series beginning with Lake Vyselki in the perch population of which the maximum number of individuals are found in the age 1+. In subsequent lakes (Table 17) it shifts to age classes 2+, 3+, 4+ and in Lake Belevets to 7+. In other words the five lakes arranged in the above series represent a single loop of change of perch population structure in most sequentially following cycles. The reason for this cyclicity as stated earlier is the influence of predation including cannibalism on the number of individuals of different age groups of fish in conditions of extremely depauperate ichthyocenoses in terms of species composition.

In view of the cyclic changes of population structure there is an attendant change in the ratio of ichthyomass of immature and mature fish. At the beginning of the cycle ichthyomass of immature (and partly mature) age groups predominates. Thus in Lake Vyselki it exceeds the ichthyomass of mature fish 14.92 times. At the end of the cycle the relationship is reversed. In Lake Belevets the ichthyomass of immature fish is only 0.01 of that of mature fish.

In perch populations from perch-pike lakes mass maturation is observed in the age classes 3+–4+, that is, in this index the investigated populations are similar to the ones investigated from lakes with multispecies fish population and closer to that from perch lakes.

The correlation of the age of ichthyomass culmination and mass maturation in populations from perch-pike lakes is similar to that observed in populations from pure perch lakes. In three cases this age is the same; in one, ichthyomass culmination precedes the age of mass maturation of fish and in one succeeds it. On the basis of these data as also the ratio of ichthyomass of immature and mature age groups and other indices examined above perch population from perch-pike lakes belong to the third and second types.

Considering the entire gamut of data on the ecology of perch from perch-pike lakes it may be thought that the structure of its populations changes cyclically similar to the dynamics described for perch lakes. Herein lies the basic similarity between perch populations from the above two types of reservoirs.

In concluding this section it is necessary to dwell on the following methodological question. Interesting data on the structure of perch populations from perch and perch-pike lakes having general ecological importance were obtained, thanks only to the uses of ichthyocides. During use of casting nets even as modified for the total catch from reservoirs [Zonov, 1974] a doubt always remained about the reliability of data on the estimates of younger age groups of fish whose numbers in the perch lakes change very rarely. The deep-seated distrust of the data on estimates of fish with the help of casting nets (as well as trawls) made it extremely difficult to select the initial premise for subsequent analysis (including construction of a mathematical model) of the dynamics of perch population from small lakes inhabited by only this fish species. However, a comparison of the methods of direct estimates of fish by the area method with estimates after the ichthyocide treatment of lakes allows us to think that the data obtained during catches using larger fine-mesh casting nets give reliable information on the number of fish and the nature of structure of their population.

DISCUSSION

The material presented above relating to the structure of populations of some species of fish from small lakes and its dynamics makes it possible to put forth several propositions about the specific reservoirs under investigation and also, in our opinion, offer more generalizations. It is hoped that this is in fact so inasmuch as the estimates of fish in small lakes through catches with large fine-mesh catchnets and treatment of reservoirs with ichthyocides make it possible to describe the basic features of the structure of fish populations in a fairly complete and objective manner. It is important that the method employed for the purpose facilitates a discussion of all the problems covered by the scope of the investigations at the qualitative level, which more adequately reflects the essence of population structure than do qualitative descriptions.

To analyse the structure of populations, wherever possible and permissible by the initial data, such index as ichthyomass was used. The number of individuals was, as a rule, used as an additional criterion, and also, in those cases where it reflected the essence of the phenomenon. The accent on the use of ichthyomass as an important index in the analysis of the structure of populations imparts to the data obtained from a study of a limited number of reservoirs —small lakes—a more general significance. This is not incidental but is because ichthyomass is to a greater extent than the number of individuals, closer to the rank of integral indices and in view of this, more completely reflects the deep-seated populational phenomena*. A higher status of ichthyomass as one

---

* The most completely generalized description of populational processes in relation to general ecological phenomena is possible at the fish production level. However such investigations as yet are few and far apart and in essence have only begun to grow. As an example I cite the following works: [Rudenko, 1967; Rudenko and Volkov, 1974; Umnov and Rudenko, 1979 ; Studies on Inter-relationships . . ., 1986; Rudenko, Kuderskii and Pechnikov, 1988 and others].

of the population-structure indices allows us to recommend its wide use in the practical studies on fishing as an initial index and not as one deduced through the number of individuals.

The generalized character of the situations obtained above from the example of fish populations from small lakes imparts another feature to it, that is to say, these are based on absolute and not relative values of the initial parameters. And even if in the determination of ichthyomass and the number of individuals certain errors are permissible within the error limits of the methods used, all the same these describe more adequately the peculiarities of the structure of fish populations than do widely used relative values obtained from the analysis of the average samples. Thanks to the great correspondence of absolute indices with the actual course of the populational phenomena, the conclusions arrived on their basis although they relate to the investigated small lakes are applicable to populations from different types of reservoirs.

In view of the above said it is entirely expedient to specially examine some general propositions flowing from the actual data on individual species of fish presented above. To facilitate applicability of the material in the text that follows, the descriptions has accent on particular chosen problems.

Range of Age Series in Fish Populations

Such an index of the structure of population as the length of the age series in the investigated seven species of fish from small lakes is subject to great variation. The limiting length of the age series, expressed by the maximum age of fish in a population, varies in five species 2.0–3.5 times (Table 18). In white bream and bleak the differences are less but it is quite possible that they are associated with the limited number of investigated populations of these species.

Table 18. Range of age series in fish populations from small reservoirs

Species	Limiting age of fish in a population			Number of populations
	Lowest	Highest	Ratio	
Roach	7+	15+	2.1	13
Bleak	6+	7+	1.2	5
White bream	7+	10	1.4	2
Bream	7–7+	20–20+	2.9	45
Ruff	4+	8+	2.0	9
Pike	4+	14+	3.5	14
Perch-1	6+	15+	2.5	13
Perch-2	5+	12+	2.4	10

Note: For perch populations it is conventional to divide them in two groups: from lakes with multispecies fish population (1) and from perch and perch-pike lakes (2).

A considerable amount of variation in the range of age series is not something unexpected. This fact is well known in ichthyological literature. The essence of the problem boils down to the factors responsible for such variable limits on the age of fish within the same species but from different populations. Often a shortened age series in one population in comparison to the other is explained either by the limited data collection or by inadequate trappings by the fishing gear. In both cases the fate is the same: it is confirmed that during investigations, the older fish are not observed. Often this phenomenon is linked to the effect of fishing. As applied to the investigated fish populations from small lakes all these arguments basically fall flat. As pointed before, fishing in most of the investigated reservoirs was either not intensive or (most often), lacking altogether. Many lakes treated with ichthyocides, including all perch and perch-pike lakes, do not have any fishery importance. If we speak about the effect of fishing, in recent times in most small lakes it has led to the so-called neglect resulting in a higher density of fish population. Thus in small lakes of Pskov Region during 1948–1959 on the average the annual catch was 1250 ton, during 1981–1985, 370 ton or 5/17 of the original [Perminov and Aleksandrov, 1986].

Hence, explaining the shortening of age series in fish populations from small lakes as due to the effect of fishing is the simplest of ways out of the situation. But what would happen if fishing were absent and yet there are considerable changes in the age structure?

The vast material examined for the structure of many fish populations does not support the effect of random neglect in consideration of individual specimens of fish for the range of age series. Moreover it is necessary to bear in mind that in the fish-free reservoirs consequent to the use of ichthyocides the older age groups of fish were considered most completely. Hence it must be conceded that the described changes (most often shortening) of the range of the age series in fish populations are basically due to natural causes or ecological factors. An explanation has gained favor, that the entire gamut of known, and to a large extent, unknown causes, or factors are clustered under the neutral concept "natural mortality". I have no basis to refrain from using this concept but at the same time, cannot verify its specific content.

Perch populations are most demonstrative for the purpose of the problem under discussion. The range of the age series in these populations in the pure perch lakes is 6+–9+, in perch-pike lakes it is 5+–12+ and in lakes with multispecies ichthyofauna 6+–15+. The effect of natural mortality is there, but it may be considered that in each reservoir it is manifest in a specific manner. No less interesting are the data on the populations of bream, among which despite the neglect of fisheries in 6.7% cases the age series breaks at classes 7–8+, still in 11.1% at 9–10+, although in 51.5% populations the age series extends upto 15+ and more. Here too one sees the effect of natural mortality attained against the background of intrareservoir ecological processes. Thus the observed variation in the range of age series in fish populations from small lakes occurs mostly under the impact of natural mortality, for which a detailed study of its concrete forms in the individual reservoir is indicated.

Decrease in the Number of Individuals and Ichthyomass with Age in Fish Populations

The number of individuals in age classes in a fish population decreases continuously starting from 0+. As opposed to the number of individuals the ichthyomass increases for several years in a row but then it too decreases. As the estimates show, the rate of decrease in the number of individuals* in all the investigated species of fish is higher than that of ichthyomass. According to the data for some populations the former constitutes 1.9–3.6 (average 2.7) while the latter is 1.1–22.8 (average 1.8). The average rate of decrease of ichthyomass of age groups as fish become old is roughly 2/3 of that of the number of individuals. The slow rate of ichthyomass decrease is due to the continuous weight gain by fish which does not cease until they die.

In a population, the rate of decrease of ichthyomass (like the number of individuals) changes nonuniformly with age; firstly, in view of the peculiarities of mass growth of the population and secondly, due to differences in the ichthyomass of various generations, including mixed ones.

The rates of decrease of the number of individuals and ichthyomass in populations of different species of fish are not identical. These are the highest in perch from perch and perch-pike lakes and lowest in pike and bream. These are likewise high in perch from lakes with multispecies fish population, bleak, roach and ruff. In individual populations of one and the same species the rate of decrease in the number of individuals and ichthyomass in age groups usually do not coincide although the character common for the species is always retained.

A comparative analysis of the rates of decrease in the number of individuals, and particularly, ichthyomass in age groups may prove of great importance in a detailed study of the fish production phenomena not only at the population but also at the ecosystem level.

---

* The rate of decrease in the number of individuals was calculated as the ratio of the number of individuals of the succeeding age groups to the preceding. The rate of decrease of ichthyomass was also calculated using the same principle.

Age of Attaining Mass Maturity

The process of maturation of fish in populations, as a rule, stretches over many years. Moreover, its rate is different in males and females. While describing individual species of fish in preceding sections the emphasis was on the rate of maturation of females since the reproductive potential of population is related to the maturation of its females. In the following discussion too, the account pertains to females.

The age of ónset of mass maturation in most investigated species (six out of seven) from small reservoirs fluctuates within the range of two–three age classes. Thus mass maturation in roach, ruff, pike and perch from lakes with multispecies ichthyofauna is observed in the age of 3+–4+ (or 5 completed years). In bleak and white bream it is respectively 4 and 5–5+ while in perch from perch and perch-pike lakes it is 2+–4+. Unlike these species, in bream the age of attaining mass maturity changes in individual populations from 3–3+ to 8–8+ or in the range of six age classes.

In almost all species of fish the number of mature groups is variable, that is, the number of age classes following the year of attaining mass maturity. It varies from three in bleak to 11 in roach, 12 in perch from lakes with multispecies fish population and 13 in bream (Table 19). It is customary to consider that populations possessing a prolonged age series, including in addition a large number of mature age groups of fish successfully counter the external influences including those in the nature of a stress factor. In such populations the structure is more stable (in a multiyear section) and the reserves too are more stable. However in Table 19 what draws our attention is not the range of the age series in individual populations but a considerable variation of this index when we examine the entire totality of populations of one species. Thus in bleak populations two–three mature age groups older than fish reaching mass maturity and in roach from 2 to 11 are noticed. In populations of perch from lakes with multispecies ichthyofauna 1–12 mature groups of fish are found while in bream it is from 3 to 13 and in ruff 2 to 5.

In species of fish with prolonged mature age series the extreme versions are considered not as exceptions but as entirely normal phenomena. Actually in roach 23.1% populations have only two–three mature age groups (except those that reached mass maturity), 30.8% have five–six, 30.8% seven–eight and 15.3% nine–eleven age groups. In bream 9.4% populations have three–four mature age groups, 21.9% have six–seven, 25.0% have eight—nine, 28.1% 9 to 11, and 15.6% 12 to 13. Despite such a high variation in the number of mature age classes of fish in individual populations none of them can be considered in the state of decline or disappearance.

Table 19. Distribution of number of populations of different species of fish from small
lakes according to the range of the age series (in years)
after the age of attaining mass maturity

Fish Species	Range (years) of age series after the age of onset of mass maturation													Number of populations
	1	2	3	4	5	6	7	8	9	10	11	12	13	
Roach	—	2	1	—	1	3	1	3	1	—	1	—	—	13
Bleak	—	3	2	—	—	—	—	—	—	—	—	—	—	5
White bream	—	1	—	—	1	—	—	—	—	—	—	—	—	2
Bream	—	—	1	2	—	3	4	4	4	6	3	3	2	32
Ruff	—	1	2	5	1	—	—	—	—	—	—	—	—	9
Pike	1	1	6	1	1	1	1	—	1	—	—	—	—	13
Perch-I	2	—	2	—	2	2	1	—	2	1	—	1	—	13
Perch-II	—	2	2	1	4	—	—	—	1	—	—	—	—	10

Note: For perch populations it is customary to divide them in two groups: from lakes with
multispecies fish population (1) and from perch and perch-pike lakes (2)

The material available at my disposal, unfortunately, does not permit an analysis of the peculiarities of ecological environment of reservoirs, the fish populations from which are presented in Table 19. One can with certainty only say that the investigated species of fish successfully exist and satisfactorily reproduce stocks in concrete conditions of lakes inhabited by them. The above described situation poses many questions on the problem of the dynamics of fish populations and the factors responsible for it, the role of the peculiarities of structure and stability of population and the magnitude of ichthyomass constituting it and so on. Obviously, despite the usual understanding, these problems are more unique and complex.

Decrease in the Number of Individuals and Ichthyomass of Age Classes of Fish by the Time of Onset of Mass Maturation

The age of mass maturation of fish is an important population structure index. Moreover it is a boundary separating two qualitatively different parts of the population: immature and mature*. One of them relies on the age of mass maturation in order to develop the exploitation regime of commercial stocks of fish. Hence it is quite natural to compare with it the different aspects of intrapopulational processes including the change of the number of individuals and ichthyomass.

---

* In age classes directly preceding mass maturation there are some individuals that have attained maturity precisely as in groups of fish a large part of which have matured (mass maturation) there are immature individuals. Considering the general nature of the present book such a detailed division of immature and mature fish is not attempted.

Table 20. Decrease in the number of individuals and ichthyomass of age classes in fish populations from small reservoirs by the age of mass maturation

| Fish species | Decrease (times) | | | | Number of populations |
| | Number of individual | | Ichthyomass | | |
	Range	Average	Range	Average	
Roach	15.4–186.7	63.2	1.6–25.6	6.2	12
Bleak	3.1–129.2	35.1	1.6–9.2	3.2	5
White bream	8.4–40.0	24.2	1.9–3.0	2.9	2
Bream-a	1.6–170.0	29.2	1.1–5.2	2.3	14
Bream-b	1.6–170.0	29.2	1.1–10.6	2.3	10
Ruff	9.7–227.4	44.3	1.4–7.5	3.5	9
Pike-a	3.7–104.2	25.3	1.01–3.3	2.0	6
Pike-b	14	14	1.2	1.2	1
Perch-1	20.5–391.7	155.6	1.1–33.4	9.9	13
Perch-2	2.8–16.6	6.8	2.3–3.4	2.2	3
Perch-2	2.2–4.4	3.3	3.1–5.4	4.3	2

Note : (1) For perch populations, it is customary to divide them in two groups; from lakes with multispecies population (1) and from perch lakes (2).

(2) For bream, pike and perch-2 the second series of figures are given for populations belonging to the third type.

Unlike the number of individuals, the decrease in ichthyomass is not that considerable, which is connected with the continuous weight gain by individual fish. The maximum decrease in ichthyomass is observed in individual roach and perch populations and reaches 5/128–5/167* of the original (on the average 5/31–10/99).

The data presented for the decrease in ichthyomass relate to populations of the third type. In populations of the second type, in which ichthyomass culmination occurs after fish attained mass maturity, the age related dynamics of this index is somewhat different. In bream populations of the second type the ichthyomass of age classes from the moment of mass maturation to its culmination increases 1.1–10.6 times; in pike 1.2 times; in perch from perch lakes 3.1–5.4 times. In other words if in populations of the third type the ichthyomass of age groups decreases continuously from culmination to mass maturation of fish, then in populations of the second type its value rises and attains the maximum (culmination) after the onset of mass maturation. These differences are of great importance for establishing the regimes of rational exploitation of commercial fish stocks in specific populations.

---

* While determining the degree of decrease of ichthyomass by the age of mass maturation of fish the value of ichthyomass culmination was taken as the basis.

Among the data presented here, of great interest is the wide range of variation of the value of ichthyomass during a changeover from the age of its culmination to mass maturation in populations of the third type and the reverse transition in populations of the second type. This variation confirms the great qualitative uniqueness of biotic processes in ecosystem of various small lakes.

The ratio of ichthyomass of immature and mature fish is intimately connected with the change of value of ichthyomass of age groups from its culmination to mass maturation. In such species as roach, bleak, white bream, ruff and perch from lakes with multispecies ichthyofauna the major part of the ichthyomass of population is contributed by the immature (and partly mature) age groups (Table 21). This contribution is quite high in roach and perch in whose populations the immature fish contribute to ichthyomass on the average 9.1–10.4 times that of mature ones. All such populations of the above five species of the fish belong to the third type.

Populations of bream, pike and perch from perch lakes in the ratio of ichthyomass of immature to mature age classes are not homogeneous, and can be split into two groups. In one, as in the five examined species of fish, the total value of ichthyomass is more in immature fish;

Table 21. Ratio of ichthyomass of immature to mature age groups in fish populations from small lakes

Fish species	Ratio of ichthyomass of immature to mature age groups		Number of populations
	Range	Average	
Roach	2.5–29.6	9.1	12
Bleak	1.5–14.0	5.0	5
White bream	4.0–7.3	5.7	2
Bream-a	1.1–4.6	1.9	16
Bream-b	0.04–0.8	0.4	16
Ruff	1.2–7.2	3.3	9
Pike-a	1.1–9.4	3.6	4
Pike-b	0.1–0.9	0.7	10
Perch-1	1.9–35.8	10.4	13
Perch-2a	1.3–2.5	1.9	2
Perch-2b	0.04–0.5	0.2	3

Notes: (1) For perch populations, it is customary to divide them into two groups; from lakes with multispecies population (1) and from perch lakes (2).

(2) In bream, pike and perch-2 populations have been identified with predominance of ichthyomass of immature age groups (a), and predominance of ichthyomass of mature age groups (b).

in the other it is the mature fish. However the magnitude of the ratio of ichthyomass of these two groups of fish in bream, pike and perch from perch lakes is lower than in roach, bleak, white bream, ruff and perch from perch lakes with multispecies fish population. This confirms the lesser contribution of the immature part of the population in the first three species.

In the second group of populations of bream, pike and perch from perch lakes the total ichthyomass of immature age groups is less than that of the mature age groups. On the average the former is 0.2–0.7 of the latter.

The higher ratio (more than 1) of ichthyomass of immature to mature fish in populations of bream, pike and perch from perch lakes is not identical. This happens because a part of the populations of these fish belongs to the third type and the other to the second type.

It must be noted that in works on fishery evaluation of small lakes the differences between individual species of fish and their populations in respect of ichthyomass of mature and immature fish, as revealed by the data presented in Table 21, are not taken into account. Such an approach can never be considered correct. Heterogeneity of populations of different species of fish in this respect is reflected in the variation of the production processes which has a great significance in the general limnological and fishery evaluation of lakes.

Age of Ichthyomass Culmination

In ichthyological, fisheries and fish production studies this index is almost not used. Only a few authors may be named who have sometimes used the term "ichthyomass culmination" in their works (for instance Tyurin, 1963). Such a treatment of the important population structure index is rather unjust because it is of principal significance, for example, for an objective assessment of the fairly complex problems of regulating fishing and completeness of utilization of resources of reservoirs [Kuderskii, 1983, 1984, 1986].

Ichthyomass culmination in populations of different species of fish and in different populations of one and the same species sets in not simultaneously. This is due to heterotypic dynamics of weight gain of fish. According to the material from small lakes the earliest age of ichthyomass culmination (0+) is observed in populations of roach, bleak, ruff and perch from reservoirs with multispecies fish population (Table 22). Moreover ichthyomass culmination in the age class 0+ in the above cited species is neither incidental nor an exception. For roach such a phenomenon was noted in 41.7% of investigated populations, in bleak in 25.0%, in ruff in 55.6% and in perch from lakes with multispecies

Table 22. Age of ichthyomass culmination in fish populations from small lakes and the time gap between it and the age of mass maturation of fish

Fish species	Age of ichthyomass culmination	Gap in years between ichthyomass culmination and mass maturation		Number of populations
		Range	Average	
Roach	0+–1+	2–5	3.7	12
Bleak	0+–3+	1–4	1.8	5
White bream	2–4	1–3	2.0	2
Bream-a	2–2+–5–5+	1–5	2.4	14
Bream-b	5–5+–11–11+	1–5	2.0	10
Ruff	0+–1+	2–4	2.7	9
Pike-a	1+–5+	1–4	2.0	7
Pike-b	5+	1	1.0	1
Perch-1	0+–1+	2–5	3.4	13
Perch-2a	2+–3+	1–2	2.3	3
Perch-2b	6+–7+	1–3	6.5	2

Notes: (1) For perch populations it is customary to divide them into 2 groups: from lakes with multispecies fish population (1) and from perch lakes (2).

(2) In bream, pike and perch-2 we have identified populations of the third type (a) and second type (b).

composition of fish population in 38.5%. In the remaining species of fish ichthyomass culmination occurs in age classes 2+ and older*. The maximum age of ichthyomass culmination is observed in pike (5+), perch from perch lakes (6+–7+) and bream (11–11+)

The change of the age of ichthyomass culmination in different populations of one and the same species is due to firstly the harvest capacity of individual generations and secondly due to the intensity of elimination of the young ones. The effect of the latter factor is especially demonstrative in the example of dynamics of the structure of populations of perch from perch and perch-pike lakes which are entirely governed by the "predator-prey" factor.

In bream and pike the limiting age of ichthyomass culmination is determined by the fact that both these fishes belong to species with prolonged life cycle. Hence with the appearance of the high harvest generation it may retain the position of the culminating group for many years in a row. In pike from the investigated small lakes this phenomenon is observed up to the age of 5+, in bream 11–11+.

The age of ichthyomass culmination is different in population of one and the same species of fish belonging to the third and second types. In the latter it is more than or in an extreme case coincides with the

---

* In one out of 14 populations ichthyomass culmination was observed in the age 1+.

maximum value for populations of the third type (Perch, bream and pike).

Thus the age of ichthyomass culmination is not something incidental. It depends firstly in general terms on the peculiarities of each species (short life cycle, long life cycle) and secondly on ecological environment established in a reservoir. In turn ecological factors become manifest through the magnitude of harvesting capacity of individual generations and the intensity of elimination which is dependent on the food availability to different age (and size) groups of fish and the effect of "predator-prey" type relationships. In individual cases such relationships as "parasite-host" could play a substantial role in the elimination of fish [Pronin and Khokhloa, 1986].

The role of ecological factors in many cases becomes decisive for determining the concrete period of ichthyomass culmination in individual populations. Hence the important task of future investigations is to examine this phenomenon in detail in relation to the dynamics of ecological environment in the reservoirs.

Correlation of the Age of Ichthyomass Culmination and Mass Maturation of Fish

In fish populations from lakes, ichthyomass culmination coincides with the age of mass maturation of fish, succeeds it or precedes it. The third case is the most common. Of the seven species of fish investigated, in all populations of roach, bleak, white bream, ruff and perch from lakes with multispecies fish population, ichthyomass culmination sets in before mass maturation (Table 22). All these populations belong to the third type. A more detailed analysis reveals that in addition in many populations of roach, ruff and perch from lakes with multispecies ichthyofauna ichthyomass culmination is obtained even before the appearance of first matured individuals.

Unlike the above listed species, in bream, pike and perch from perch and perch-pike lakes the correlation of ages of ichthyomass culmination and mass maturation is different. Several populations of these species belong to the third type and ichthyomass culmination in them precedes the mass maturation. Besides, some populations have second type of structure. Ichthyomass culmination in such cases sets in after attaining mass maturity.

The position of age classes according to ichthyomass culmination before or after mass maturation in population of bream and pike is determined mostly by the harvesting capacity of individual generations and the degree of elimination of the young ones. One may observe such combination of these two factors at which the high harvest generation remains in the position of ichthyomass culmination even after attaining mass maturation. However such a situation is not very commonly observed. Out of the 32 populations of bream investigated it was observed in 10 (31.3%) whereas in 14 (43.8%) culminating age class had young fish in the state of mass maturation. For pike the ratio of both states of population was still to a large extent in favor of the third type. Out of the 14 populations investigated in 7 (50.0%) the ichthyomass culmination was observed in age classes preceding mass maturation and only in one was it at an older age (Table 22). In contrast to bream and pike in populations of perch in perch and perch-pike lakes formation of structure of the third type or second type was due to the decisive influence of such biotic factor as cannibalism. Ichthyomass culminated generation crosses the age of mass maturation and becomes older than it because it intensively consumes newly born juveniles of its own species. As long as the number of individuals of the leading generation does not decrease to the critical value not one of the new generations of perch has any prospects of reaching the state of culminating ichthyomass.

The time interval (in years) between the age of ichthyomass culmination and mass maturation in populations of the third type is 1–5 years (Table 22). It is the maximum in populations of roach and perch from lakes with multispecies fish population (2–5 years). On the average, in these two species ichthyomass culmination sets in respectively 3.7 to 3.4 years before mass maturation of fish. Hence it is no surprise that in similar populations the magnitude of ichthyomass by the age of mass maturation decreases several folds (Table 20). In populations of bleak, white bream, and ruff, which also belong to the third type, ichthyomass culmination sets in earlier than mass maturation, on the average by 1.8–2.7 years.

In populations of bream, pike and perch from perch and perch-pike lakes belonging to the third type the difference between the age of ichthyomass culmination and mass maturation also appears to be at the level reported for bleak, white bream and ruff and on the average is about 2.0–2.4 years. However in bream, pike and perch from perch and perch-pike lakes the age gap is found also in populations belonging to the second type. In this case it is on the average 2.0 years in bream, and even up to 6.5 years in perch. For pike only one population of type two was investigated. The age interval for this (one year) between the two events in question should not be considered indicative.

No attention was paid to the above stated gap between the onset of ichthyomass culmination and mass maturation of fish in investigations on the dynamics of stocks of commercial fish since the very concept of ichthyomass culmination as a rule, was not analyzed. However as will be seen from the data presented here, the difference between the age of onset of each of these two phenomena has to be taken into account for the fishery and fish production evaluations of fish population as also for developing regimes of rational exploitation of live resources. It is also absolutely clear that the very possibility of estimating such population-structure indices as the age of ichthyomass culmination and the difference between it and the age of mass maturation arises only when one uses the methods of direct determination of the number of individuals and ichthyomass of fish in reservoirs in place of methods of their relative estimates. Hence a transition from investigations in applied ichthyology and fish production phenomena to a new level of demands faster introduction of various methods of fish estimation in reservoirs such as direct conventional methods as well as ones based on distance principles (acoustic methods and so on).

Effect of Fishing on the Population Structure of Fish from Small Lakes

The material on the structure of fish populations from small lakes discussed above was obtained in the absence of any regular fishery in these lakes. Only in some lakes low intensity casual catches were conducted. However the problem of the possible influence of commercial catches in small lakes on the structure of fish populations retains its urgency. Hence it was thought expedient to elaborate it briefly on the example of the small lake Pelyuga (106 ha) where intensive catches were conducted for a period of 11 years (1974–1984) using large fine-mesh catchnets*. The influence of catches also on the population-structure indices is evaluated such as range of the age series, ichthyomass culmination and its age correlation with the onset of mass maturation of fish.

It is necessary to mention here the types of fishing being practised. Based on the task of the problem at hand fishing must be divided into less intensive, and intensive destructive. The last type of fishing leads to the so-called overfishing [Chugunov, 1928; Ressell, 1947] and is not considered here. Intensive fishing is of interest to us, that is, such a pressure of utilization of reserves at which the per cent reduction of ichthyomass on the average for many years does not surpass the coefficient of natural mortality in the caught fish. It is such type of fishing that is carried out in Lake Pelyuga. Over a period of 11 years in it 8.0 to 66.3% of ichthyomass was annually removed. On an average for the entire period of the work, catches were of the order of 55.7 kg/ha or 26.1% of the mean annual ichthyomass equal to 213.4 kg/ha. Without

---

* Kuderskii, Pechnikov and Rudenko (1988) have discussed in detail the material relating to this problem.

going into details it must be mentioned that such fishing does not lead to destruction or serious restructuring of populations of the caught fish to which (in Lake Pelyuga) belong ruff, roach, bleak, perch and pike. For the first four species the intensity of catch at the level of 26.1% of the ichthyomass is below the fishing "load" which may cause a serious deformation of the structure of its population. Nevertheless fishing in Lake Pelyuga being carried out over years with variable intensity had led to a variation in the range of age series of the species of fish in question (Table 23).

After the first three years (1974–1976) of commercial exploitation the age series in populations of roach, bleak, perch and pike shortened in 1977 by 1–4 age classes. On the other hand in ruff in 1977–1978 the age series extended by 1–2 classes but later (1979–1981) it too shortened. Later there occurred a restoration of the initial age structure of the population. Despite the fact that in 1980–1982 the intensity of fishing was higher as compared to preceding years, by 1982 there was prolongation of the age series in all species of fish in relation to their state during the period 1977–1980. In pike and bleak the number of age classes already restored in 1981, in ruff in 1982 and in perch in 1984. Only in roach was the range of age series up to the last year of observation shorter than the initial by one class. In the ultimate analysis, despite years of intensive fishing the number of age classes in populations of fish was equal or almost so to the initial value and in pike even exceeded by one class. However, it is characteristic that the dynamics of the number of age classes in fish populations existing for a period of 11 years does not show a direct correlation with the percentage decrease of ichthyomass as a result of intensive fishing. It could be suggested that shortening of the age series in the middle of the period of investigation (1977–1984) occurred due to relatively faster depletion of the lower harvest generations predominating in the first years of conducting intensive fishing. However, already during the process of such fishing in Lake Pelyuga, productive and high harvest generations appeared in the species of fish under consideration because of which the initial age structure of populations was restored (or almost restored).

The effect of fishing intensity on the age of ichthyomass culmination is quite apparent from the data presented in Table 24. In populations of pike, perch and bleak the perennial intensive fishing does not affect the age of ichthyomass culmination. Another phenomenon is observed in ruff and roach. For them the characteristic increase of the range of age classes at which ichthyomass culmination occurs is 0+–5+. A similar analysis of the problem on the example of ruff is available in the work of Kuderskii, Pechnikov and Rudenko (1988).

Table 23. Limiting age in populations of fish from Lake Pelyuga

Fish species	Year of fishing										
	1974	1975	1976	1977	1978	1979	1980	1981	1982	1983	1984
Ruff	6+	6+	6+	7+	8+	4+	4+	5+	6+	6+	4+
Roach	10+	10+	10+	7+	9+	7+	6+	6+	7+	8+	9+
Bleak	6+	6+	6+	5+	5+	5+	5+	6+	7+	6+	5+
Perch	9+	9+	9+	5+	6+	7+	5+	6+	7+	8+	9+
Pike	8+	5+	6+	7+	6+	7+	7+	8+	8+	9+	9+
Catch as 5 of total ichthyomass	12.3	66.3	10.6	13.2	20.3	25.6	39.0	32.0	31.5	8.0	25.3

Table 24. Age of ichthyomass culmination in populations of fish not caught and intensively fished from small reservoirs

Fish species	Age of ichthyomass culmination in populations	
	In not caught	Intensive fishing
Ruff	0+−1+	0+−5+
Roach	0+−1+	0+−5+
Bleak	0+−3+	0+−4+
Perch	0+−3+	0+−3+
Pike	1+−5+	0+−4+

Change of the age of ichthyomass culmination in the populations of ruff and roach subject to intensive fishing is due to the restructuring of the processes of weight gain and elimination in the various age classes. In populations not subject to fishing massive elimination occurs in young age classes on account of consumption of young ones by predators, among which in small lakes the most numerous is usually the perch. The number of individuals of older fish is controlled by predators to a lesser extent and among them one observes stressed trophic relationships. Massive growth of fish in such age classes seems relatively slowed down in comparison with what is potentially possible. As a result ichthyomass culmination shifts in the numerous young age classes.

When intensive fishing is resorted to over a number of years the ratio of contributions of massive growth and elimination in the resultant ichthyomass of age classes changes. Even when fine-mesh catchnets are used the young of ruff and roach which have smaller sizes to a considerable extent "pass through" the mesh. At the same time fishing removes the older age groups of fish because of which there occurs a decrease of density and improvement of feeding conditions. The latter in turn lead to higher fecundity. As a result, howsoever paradoxical it may seem at first glance, intensive fishing of small-sized fish (perch, roach, ruff and some others) by fine-mesh catchnets leads to increasing reproductive potential of populations and a stable conservation of high number of individuals of younger age group having slower rate of weight gain. This phenomenon is well known as has been described many times [Tyurin, 1954, 1957].

In ultimate analysis the rate of weight gain of fish in the young age classes does not change or even decrease while in older age classes it increases. For example, in ruff from Lake Pelyuga after four years of intensive fishing the average weight of two-year olds either did not change or it decreased. The weight of three-year olds increased 1.5 times or (in individual years) did not change. On the other hand the average weight of individuals in the age 3+−6+ increased 2 times and in some years even more. In full agreement with such a change of rates of weight gain of ruff the age of ichthyomass culmination in its population from Lake Pelyuga shifted to the age classes 2+–3+ although in the first year of intensive fishing it was 1+ [Kuderskii, Pechnikov and Rudenko, 1988].

On the whole, as a result of 11 year intensive fishing the age of ichthyomass culmination increased from 1+ in the first year to 3+–5+ in ruff populations and 1+–5+ in roach in the last four years.

The change of age of ichthyomass culmination could not but affect its correlation with mass maturation of fish. In ruff and roach populations from lakes not subjected to fishing or to less intensive fishing, ichthyomass culmination precedes mass maturation of fish by several years (Table 22). In these very species in Lake Pelyuga under the impact of intensive fishing for 11 years of observation the age of ichthyomass culmination in ruff populations succeeded the time of onset of mass maturation by two years and in case of roach by one year. Moreover both these phenomena coincided in time in populations of ruff for four years, and in roach for one year. As a result ichthyomass culmination preceded mass maturation in population of ruff for 5 years and in roach for nine years. In other words the structure of population of ruff and perch is of third type typical of lakes not subjected to fishing. In Lake Pelyuga in individual years it changed over to the second type.

The data examined here do not confirm the conventional ideas about the effect of intensive fishing on the structure of populations of fish.

Periodic Changes in the Age of Ichthyomass Culmination in Fish Populations from Small Lakes

In a population the affinity of ichthyomass culmination to any age class is determined by several factors among which we can list the magnitude of harvest of individual generations, pressure of natural elimination in various age groups in unexploited and weakly exploited lakes and the effect of intensive fishing. The degree of manifestation of these and some other factors is different in different reservoirs and could lead to situations in which there is a periodicity of their combined action on fish populations. In such cases one cannot rule out manifestation of periodicity also in the change of the age of ichthyomass culmination. This phenomenon has been described earlier on the example of the bream and perch (Figs. 6 and 8).

In populations of bream from unexploited small lakes other factors being equal for several years periodicity of change of age of ichthyomass culmination is determined by the level of harvest. When there is a generation marked by the level of harvest among neighboring generations it, at not very high elimination, may occupy the leading position in respect of ichthyomass and retain this position for several years on end. In such a case, ichthyomass culmination for several years appears to be linked with the definitive position and age of culminating group of fish and increases as this particular generation becomes old (Fig. 6). With the appearance, after several years, of the next generation also marked by the level of harvest, the process repeats and in bream populations there occurs a periodicity of change of the age of ichthyomass culmination.

For examining such a periodicity, of great interest is the change of correlation of the maximum age of ichthyomass culminating group of fish with mass maturation. Among populations of bream from small lakes often one observes a situation in which the maximum age of the culminating group of fish in the entire period of its change remains less than the age of onset of mass maturation. In such a case the structure of bream population in all years, during which one generation retains the leading position, is similar to the third type.

The high-harvest generation, unlike the common one, may be found among other generations at the leading position for a longer period of time. Hence other conditions remaining equal it sometimes retains the leading position also after attaining mass maturation. In this case the age of ichthyomass culmination is higher than the age of onset of mass maturation in fish and the population structure corresponds to the second type.

In populations of perch from perch and perch-pike lakes the periodicity of change of the age of ichthyomass culmination is governed by "predator-prey" type relationships. The duration of each period is determined by the rapidity of elimination of the leading (in ichthyomass) group of ichthyophagous fish under the impact of natural mortality.

Differences in the nature of periodicity of change of the age of ichthyomass culmination in bream and perch populations are quite obvious. However these differences lead not alone to factors determining the emergence of periodicity. In the bream populations the periodicity in question may or may not be present since the appearance of fairly numerous generations in a strict sequence, against the background of allied populations, is not strictly regulated. In perch populations the situation is otherwise. In them the periodicity of changes of the age of ichthyomass culmination is orderly and in essence it should be viewed as a cyclic process of restructuring of the population. The orderly nature of this cyclicity accrues from the specific biotic relationships operating within the ecosystem of perch and perch-pike lakes. Only the duration of the cycle may prolong or shorten in some years but the process per se is conserved.

Both examples of periodic changes of the age of ichthyomass culmination are in the nature of logically complete schemes. For bream the scheme is supplemented by a set of possible states of the structure of population based on the data of different lakes (Fig. 6). The example of perch is supplemented by a set of states of the mathematical model describing the dynamics of structure of population of this species from perch lakes (Fig. 8). However in the degree of clarity and correspondence with the actual dynamics of structure of fish populations both examples are inferior to the data of direct natural observations. Such material is available for Lake Pelyuga in which continuous intensive fishing and observations on the structure of populations of ruff, roach, bleak, perch and pike were conducted for a period of 11 years (Kuderskii, Pechnikov and Rudenko, 1988]. The data obtained on the dynamics of the age of ichthyomass culmination are presented in Table 25. In it the changes of the age of ichthyomass culmination are identified as periods in those cases where one and the same generation retained the leading position in the population for three years and more.

Out of the five fish species investigated here, in pike for the entire duration of 11 years of observations the periodicity in the change of the age of ichthyomass culmination was not observed. In populations of ruff, roach and perch the periodicity was not constant in character; it either discontinued after two periods (ruff, perch) or alternated for several years disturbing the orderly change of the age of ichthyomass culmination (roach) necessary for retaining the periodicity. Only in the populations of bleak over the entire period of 11 years there occurred sequential replacement of one period by the other, which is linked to the constant appearance, after some years, of generations that are more

Table 25. Age of ichthyomass culmination in fish populations from Lake Pelyuga in different years of fishing

Fish species	Year of fishing										
	1974	1975	1976	1977	1978	1979	1980	1981	1982	1983	1984
Ruff	1+	3+	0+	1+	2+	3+	2+	3+	4+	5+	3+
Roach	1+	2+	3+	1+	1+	1+	2+	3+	4+	5+	1+
Bleak	1+	2+	3+	2+	3+	4+	1+	2+	3+	0+	1+
Perch	0+	1+	2+	0+	1+	2+	2+	2+	2+	2+	3+
Pike	1+	3+	3+	4+	3+	1+	1+	2+	4+	0+	1+

Note: The periods of change of the age of ichthyomass culmination have been identified with an underline. Age classes which have reached mass maturation have been shown in box.

numerous than the neighboring ones. In all in bleak populations three periods changed and in 1983–1984 the fourth period just began.

In populations of ruff and roach there was a clear manifestation of the effect of the magnitude of harvest of generation on the length of the period of change of the age of ichthyomass culmination. In 1977 generation of ruff was 2.6 times as numerous as the preceding and 1.4 and 5.1 times the two succeeding ones. As a result it retained the leading position for a period of four years (1980–1983), having the age respectively from 2+ to 5+ and crossed the limit of mass maturation. Because of this in the generations of two years (1982 and 1983) the age of ichthyomass culmination was higher than the age of mass maturation. In roach populations the 1978 generation was 1.3 times more numerous than the previous generation and 2.5 to 5.3 times respectively of the two succeeding generations. Thanks to this and other conditions remaining the same, it occupied the leading position for five years (1979–1983) having increased the age from 1+ to 5+. This roach generation crossed the age of mass maturation which occurred only once (as in ruff) in 11 years of observation. In view of this ichthyomass culmination in 1983 in the roach population was observed in the age class older than mass maturation of fish.

Generalizing the material on fish populations from Lake Pelyuga it must be mentioned that the periodicity of change in the age of ichthyomass culmination is not a rule. It arises under a favorable culmination of biotic and abiotic environmental factors which is not always possible. In conditions of Lake Pelyuga its manifestation was facilitated by the intensive fishing which had its impact on the variation in the age of ichthyomass culmination in population of ruff and roach as also due to a change in the degree of elimination of various age groups of fish.

The extensive material presented in this chapter relating to the populations of seven species of fish from small lakes clearly indicates the scope of complexity of the problem under discussion. At the same time it illustrates expediency of analysis of the structure of populations based on the data of entire estimates of the fish populations in reservoirs and its advantage over the conventional approach based on relative data obtained by the method of average samples. It also becomes clear that for the further indepth analysis of the range of problems vast comparative material is required from many more small lakes including those belonging to different limnological types.

Fish populations from small lakes as a multicomponent formation are characterized by definite structures which are in a constantly dynamic state. The change of the structures of fish populations has a twin character. First, this structure differs in populations of one and the same fish species inhabiting different lakes. Second, the structure of a single individually taken population experiences multi-year transformation under the impact of factors external in relation to it (also in dynamic state) as also internal. The changes in the latter are directly linked with laws of functioning of the population itself.

Wide ecological and temporal variation of structure of fish populations from small lakes urgently requires methods of mathematical modeling for their analysis. So far such works are being conducted on a limited scale. Moreover their applicability for solving the problem of population dynamics of fishes from small lakes may appear productive in many respects as was demonstrated by works conducted on these lines. It is also important that the obtained solutions are relatively easily verifiable. For this relatively little effort is necessary for catches from the lakes or their treatment with ichthyocides (if the latter is possible from hygienes and other considerations).

Finally the works on the study of population-structure peculiarities of fish populations from small lakes have practical importance. In particular, their results should be utilized in developing the theoretical bases of rational utilization of production potential of these reservoirs, including the optimum regimes of exploitation of fish resources.

CHAPTER 2

STRUCTURE AND DYNAMICS OF FISH POPULATION IN RESERVOIRS IN PLAINS AND LARGE LAKES*

Ichthyological investigations in reservoirs from the plains and large lakes were conducted, as a rule, by the conventional scheme. As data base, use was made of the relative estimates of different indices, which were based on the analysis of the average samples. Such an approach made it possible to describe several aspects of the biology of fish, mostly at the organizational level.

During a changeover to the population level, the conventional scheme of investigations appears rather inadequate since it does not take into account such important indices as the number of individuals and ichthyomass of populations as a whole, and the constituent qualitatively unique groups of individuals. Only after the introduction into practice of ichthyological and fishery investigation in various reservoirs and some large lakes the methods of determination of the absolute number of fish and ichthyomass did the possibility arise to pose and discuss problems relating to the structure of populations.

In the conditions obtaining in inland reservoirs regular studies on the determination of the number of fish and their ichthyomass by the area method were initiated by I.I. Lapitskii (1962, 1967) in Tsimlyansk Reservoir. Later on, in different years, these studies were initiated in Lake Il'men' [Gulin, 1969] and several Volga reservoirs [Nebol'sina, 1980; Nikanorova, 1984]. At present, for some reservoirs and Lake Il'men', multiyear data have accumulated on the population of important commercial fish and on their ichthyomass. These data enable us to examine individual problems of the structure of populations, not only as annual

* In this chapter I have used the material contained in the following publications: Kuderskii and Nikanorov (1983); Kuderskii and Dr' nov (1984); Kuderskii, Vetkasov and Kartsev (1985); Kuderskii and Khuzeeva (1986); Kuderskii, Khuzeeva and Goncharenko (1988).

"section" but also their dynamics. From the diverse material available on the absolute determination of the number of fish, I have used original results in this chapter. Attention is focussed only on some indices of the structure of population, examined earlier with reference to fish populations from small lakes.

An acquaintance with the material discussed below does not preclude from purview the specifics of the reservoirs and the ecosystems typical of them in comparison with natural water bodies in general and small lakes in particular. Water reservoirs are not only technogenic in origin but also the dynamics of many abiotic parameters in them is artificial in nature. It is determined by the current, seldom long-term, interests of organization exploiting water resources of the reservoir. These interests, as a rule, do not correspond to the requirements of fisheries. A contradiction between fisheries and other organizations especially perceptibly influences the efficiency of natural reproduction of fish, which most directly affects the dynamics of their population structure.

Large water bodies with heterogeneous conditions of the medium and extensive shallow-water zones pose definite difficulties in works at reservoirs on estimates of fish, since they complicate trawling surveys. Hence while using the area method to determine the number of fish and ichthyomass in reservoirs errors are possible, which are obviously more than in works on small lakes.

In all large reservoirs fishing is developed. Although its intensity, as a rule, does not reach permissible limit except in Tsimlyansk Reservoir [Kuderskii and Yankovskaya, 1989], the possible influence of fishing on the population structure of some fishes cannot be ruled out. Fishing exerts specific influence on fish stocks differing in stress from the natural factors. It removes from reservoirs mostly fish which have grown to the size when it is possibly used as a food item. Elimination (and that too intensive), under the impact of natural factors, occurs mostly in younger age groups of fish.

Methods of complete determination of the number of fish and ichthyomass have not thus far been used on all reservoirs and lakes. Hence such type of data are available for a limited number of reservoirs. Below I have used material from several Volga and Tsmilyansk reservoirs and Lake Il'men' where work on absolute estimates of fish are being conducted for several years by a single method.

BREAM

In all the reservoirs under reference and in Lake Il'men', bream, a leading object of fisheries and, as a rule, over the entire period, for which data are available on the structure of populations, was the main catch. This is clear from the information given below about the bream catches and its contribution in the total catch of fish for the five-year period 1981–1985:

Reservoir	Average annual catch of bream, ton	% of total average annual catch of fish
Ivankov	171.7	69.1
Uglich	86.6	34.6
Gorkii	180.2	42.1
Cheboksar	11.7	6.2
Kuibyshev	1,679.7	37.1
Saratov	371.3	29.4
Volgograd	1,135.9	34.4
Tsimlyansk	5,060.6	39.4
Lake Il'men'	704.4	27.6

Hence, bream populations are under constant impact of fishing. However, the extent of influence of catch on the state of the bream stocks is not identical. The stock of this fish is most intensively exploited in Tsimlyansk, Volgograd and Kuibyshev reservoirs and Lake Il'men'. Very poor or low utilization of bream resources for several years is to be found in Ivankov, Uglich and Cheboksar reservoirs.

The variable intensity of fishing affects such index as the limiting range of the age series in the reservoir populations of bream (Table 26). However, this effect has a peculiar character. In those reservoirs in which the level of utilization of stocks of bream is low, the limiting range of the age series is lower than in the more intensively exploited populations. Thus, in Ivankov and Uglich reservoirs, the utilization of bream stocks is at a lower level [Nikanorov, 1984]. The stock of this fish is still less intensively exploited in Cheboksar Reservoir, which only recently has been included in the reservoirs with regular fishing [Kuderskii and Yankovskaya, 1989; Shivaev, 1986]. The range of the age series in bream populations of the first two reservoirs does not exceed 15–15+ in most years of observation (10 or 83.3%) it was 11–11+ to 13–13+. In Cheboksar Reservoir, the limiting age of bream does not exceed 16–16+. At the same time, in Volgograd Reservoir, where bream fishery is quite developed [Nebol'sina, Abramova, Ermolin and Smirnova, 1986], the limiting range of the age series in populations of this fish till the last year of investigations (1987) reached 18–18+. Moreover, the age of 16–16+ and more was observed in 8 out of 19 years (42.1%). In the intensively exploited population of bream of Tsimlyansk Reservoir, the limiting range of the age series way back in 1983–1984 was 20+ and 19+ respectively, reaching in individual years up to 22+ (1975). In 18 out of

Table 26. Number of years with limiting range of age series in bream populations from reservoirs and Lake Il'men'

Reservoir	Limiting range of age series, years								Number of years of observation
	11–11+	12–12+	13–13+	14–14+	15–15+	16–16+	17–17+	18–18+ and above	
Ivankov	3	3	4	1	1	—	—	—	12
Uglich	4	1	3	1	2	—	—	—	11
Gorkii	—	2	3	1	—	1	—	—	7
Cheboksar	—	—	1	—	2	3	—	—	6
Kuibyshev	—	1	1	3	6	—	5	—	16
Saratov	—	—	1	3	—	—	1	2	7
Volgograd	2	5	2	1	1	2	2	4	19
Tsimlyansk	2	1	1	2	2	7	3	8	26
Lake Il'men'	—	—	5	3	—	—	—	—	8

26 years (69.2%), the limiting age was 16–16+ and more. Thus intensive fishing, naturally not reaching the level of overfishing, should not obstruct the formation and conservation of bream populations with greater limiting age. At the same time, low-intensity fishing may entail a sharp contraction in the age series together with deterioration of such biological indices as the weight gain and body length at which the fish attain maturity.

It follows from the above-said that the widely prevalent view about the catastrophic contraction of the age series in bream with the introduction of intensive fishing has been established under the impression of those cases in which fishing exceeded all permissible limits and reached the level of overfishing, i.e. became destructive in nature in relation to the structure of bream population.

In Lake Il'men' the range of the age series for the year of investigations (1974–1984) did not exceed 13–13+ and 14–14+. According to the published data, in mid-1960s rarely did one come across fish with age up to 17+ [Tyurin, 1972]. The age structure of bream populations, observed in 1974–1984, had a stable state despite continuously carried out intensive fishing [Tyurin, 1974; Kuderskii, Vetkasov and Kartsev, 1985].

On the whole, in the above investigated nine reservoirs, the minimum limiting range of the age series in bream populations dropped to 11–11+ while the maximum equalled 22+, that is, this index changed 2 times. In individual reservoirs, the limiting range of the age series in bream populations, in individual years, changed 1.1–2 times. The maximum value of the age series in bream, in the investigated nine reservoirs, changed 1.4 times and the minimum 1.2 times. Considerable variation in range of the age series in bream from different reservoirs, over a period of many years, confirms great adaptive possibilities of populations of investigated species in relation to this index. At the same time, one cannot forget the inconsequential, from the production point of view, presence of many older age groups because of slower growth of mass and higher coefficients.

Hence there arises the problem of optimization of age structure of bream populations in respect of specific conditions of individual reservoirs, taking into account the conditions of restoration of stocks and the food base for different age groups of fish.

The rate of decrease in ichthyomass of bream populations from reservoirs and Lake Il'men' is less than that of the population (Table 27). In view of these differences, by the age of onset of mass maturation, the number of individuals in bream populations decrease to 5/16–10/459 and ichthyomass to 5/7–5/27*. The decrease in population does not

* For obtaining comparable data for all populations the starting point was the population in age series 3–3+ and for ichhtyomass—its maximum value (ichthyomass culmination).

Table 27. Rate of decrease of the number of individuals and ichthyomass of age groups in bream populations from reservoirs and Lake Il'men'

Age	Ivankov Reservoir, average for 1976–1987		Kuibyshev Reservoir, average for 1971–1987		Lake Il'men', average for 1974–1984	
	Number	Ichthyomass	Number	Ichthyomass	Number	Ichthyomass
2+–3+	—	—	1.3	0.9	3.2	1.7
3+–4+	1.3	0.7	1.0	0.6	2.2	1.3
4+–5+	1.7	1.03	0.8	0.6	1.7	1.2
5+–6+	1.9	1.2	1.3	0.9	1.9	1.4
6+–7+	2.4	1.8	1.5	1.2	0.6	0.4
7+–8+	2.2	1.6	1.3	1.2	2.1	1.6
8+–9+	1.9	1.5	1.7	1.3	2.1	1.6
9+–10+	1.9	1.5	1.5	1.4	1.8	1.4
10+–11+	2.8	1.9	2.2	1.9	4.6	3.6
11+–12+	4.0	3.1	1.2	1.2	5.0	4.5
12+–13+	1.0	1.8	1.6	1.5	1.0	0.9
13+–14+	1.0	1.0	1.2	1.2	3.3	6.5
14+–15+	—	—	1.5	1.2	—	—
15+–16+	—	—	7.2	5.7	—	—
16+–17+	—	—	0.6	0.6	—	—

require any comments, since the theory of this process has been described but the dynamics of ichthyomass does require one. Hence it would be appropriate to dwell briefly on it. The figures reported have confirmed that by the time of mass maturation in bream, on the average of nine populations, the ichthyomass of age groups decreases to 5/16. This is quite a high value. Unfortunately, this phenomenon has not attracted due attention. But if one remembers the widely prevalent view that fishing, including even of bream, should begin only after one or two spawns, it is not difficult to understand the scale of loss of fishery due to such an approach of establishing regulated fishing.

In a superficial approach, the blame for the above reported scale of decrease of ichthyomass in the immature part of the bream populations could have been attributed to the effect of fishing. However, it is difficult to concur with such a conclusion, if we consider the following facts. First, fishing activity is regulated by fishing laws which permit catches of undersize bream in a limited quantity. Hence the main contribution of fishing to the decrease of ichthyomass in age groups is attributed to those of them the age of which is more than age of mass maturation. An exception in this respect is Lake Il'men' in which, in view of the peculiarities of the formation of bream reserves, higher rates of catches of young ones were permitted as an exception [Sechin, 1979]. Second, high indices of decrease of ichthyomass from its culmination to the age of mass maturation are observed in those populations of bream in which the reserves of this fish are used weakly. Thus, in Ivankov Reservoir, this decrease reached 10/53, in Cheboksar 10/27, and in Saratov Reservoir 10/39. At the same time, in reservoirs where bream fishery is developed, the ichthyomass of these age groups decreased: in Kuibyshev to 5/9, Volgograd to 1/23, and in Tsimlyansk Reservoir to 5/7. Third, similar indices of ichthyomass decreasing by the age of mass maturation were also established for those bream populations from unfished or weakly fished lakes which belong to the third type.

Combining the above arguments, it may be concluded that the indicated scales of decrease of ichthyomass in bream populations from reservoirs, from the age of its culmination to mass maturation of fish, are mostly due to the action of natural factors, except in Lake Il'men'. Unfortunately, in studies on the biology of bream, conducted at the above mentioned reservoirs, almost no attention was paid to the study of these factors.

The age of ichthyomass culmination in bream populations, from the averaged multiyear data, changes from 2–2+ in Lake Il'men' to 6–6+ in Uglich, Kuibyshev and Saratov reservoirs (Table 28). Moreover this index is most directly linked with the degree of utilization of reserves of this fish. Thus ichthyomass culmination in the age classes 4–4+ is observed in the weakly exploited bream populations from Ivankov and Gorkii reservoirs, and in the intensively exploited populations from Volgograd Reservoir. Precisely similar ichthyomass culmination in the age 6–6+ is observed, on the one hand, in Uglich and Saratov reservoirs and, on the other, in Kuibyshev Reservoir.

Wide range of variation of age of ichthyomass culmination becomes apparent while examining the data on structure of bream populations for each reservoir year by year. The overall range of this index increases to 1–11+. In individual reservoirs, the range of variation of the age of ichthyomass culmination varies from 2+–7+ in Volgograd and 3–6 in Ivankov, up to 4+–10+ in Kuibyshev, and 3+–11+ in Tsimlyansk reservoirs. In Lake Il'men' it is 2+–7+. On the average, for all the nine reservoirs, the age of ichthyomass culmination changes 1.8–3.7 times.

As an example of multiyear variation of the age of ichthyomass culmination, in Fig. 9 I present data from Kuibyshev Reservoir.

Considerable variation of the age of ichthyomass culmination is predominantly due to ecological factors, among which the leading role is played by the intensity of elimination of younger age groups of bream in view of the dissimilar food availability to each of them and the stress of the "predator-prey" type relationships, differences in the rates of weight gain in individual reservoirs and individual years in one reservoir, variability in the frequency of appearance of high-harvest genera-

Table 28. Number of individuals and ichthyomass of age groups in bream populations from reservoirs and Lake Il'men'

Age	Ivan'kov Reservoir, 1976–1987		Uglich Reservoir, 1977–1987		Goe'kov Reservoir, 1981–1987		Cheboksar Reservoir, 1981–1987		Kuibyshev Reservoir, 1981–1987		Saratov Reservoir, 1981–1987		Volgograd Reservoir, 1962–1987		Tsimlyask Reservoir, 1962–1987		Lake Il'men' 1974–1984	
	Million	Ton	Million	Ton	Million	Ton	Million	Ton	Million	Ton	Million	Ton	Million	Ton	Million	Ton	Million	Ton
2–2+	—	—	—	—	9.74	430	30.26	551	12.09	927	—	—	—	—	5.95	586	31.28	1,221
3–3+	8.99	462	1.61	62	4.41	462	16.48	1,039	9.22	1,031	4.90	837	12.17	1,374	10.90	1,945	5.02	422
4–4+	6.79	679	1.66	147	2.71	567	6.93	966	9.04	1,772	4.33	1,144	10.26	2,053	6.27	2,598	2.24	336
5–5+	3.92	658	2.29	345	1.84	551	3.24	891	11.58	2,809	2.12	723	7.55	2,505	4.79	3,018	1.67	354
6–6+	2.04	536	1.84	438	1.04	482	2.54	1,008	9.16	3,252	3.40	1,622	3.46	1,753	2.65	2,215	1.05	314
7–7+	0.85	300	0.83	310	0.77	469	1.43	817	6.06	2,822	2.09	479	1.97	1,347	1.86	1,880	2.11	797
8–8+	0.39	189	0.43	202	0.36	277	0.66	455	4.82	2,322	1.36	583	1.32	1,095	0.95	1,371	1.00	505
9–9+	0.21	128	0.36	227	0.16	138	0.31	387	2.85	1,841	0.46	416	0.71	690	0.78	1,067	0.49	316
10–10+	0.11	85	0.14	96	0.07	83	0.11	140	1.91	1,279	0.29	290	0.48	428	0.49	764	0.27	227
11–11+	0.04	44	0.05	43	0.04	41	0.08	112	0.86	659	0.26	289	0.23	257	0.25	438	0.05	57
12–12	0.01	14	0.03	29	0.02	25	0.05	71	0.71	573	0.18	219	0.13	225	0.19	355	0.01	13
13–13+	0.01	8	0.01	16	0.01	9	0.02	39	0.43	371	0.13	210	0.12	196	0.14	299	0.01	11
14–14+	0.01	8	—	—	—	—	0.01	16	0.36	321	0.04	72	0.04	61	0.06	164	0.004	1
15–15+	—	—	—	—	—	—	0.01	11	0.23	269	0.04	54	0.03	46	0.04	100	—	—
16–16+	—	—	—	—	—	—	—	—	0.03	47	0.02	46	0.003	27	0.04	84	—	—
17–17+	—	—	—	—	—	—	—	—	0.05	79	0.01	26	0.01	32	0.01	47	—	—
18–18+	—	—	—	—	—	—	—	—	—	—	0.01	9	—	—	0.005	18	—	—

Note: In the column "million" single underline denotes age group of fish which has reached maturity; in the column "ton" double underline denotes groups of fish with ichthyomass culmination. In Volgograd Reservoir, the age group 16–16+ includes all older ages also.

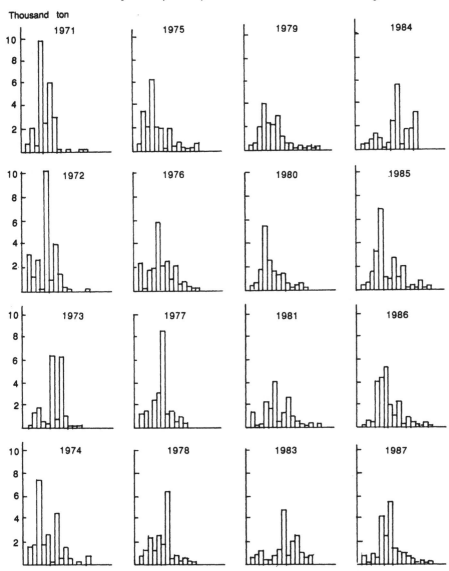

Fig. 9. Dynamics of the ichthyomass of age groups in bream populations
from Kuibyshev Reservoir.
The age class 2+ is shown to the extreme left.

tions and in their value. In individual cases (for example, Lake Il'men'),
the intensity of fishing and its stress on the depletion of relatively older
age part of the population also plays some role.

In Chapter 1, from the example of fish from small reservoirs, the
influence of the level of harvest of individual generations was proved

on the appearance of periodicity and variation in the age of ichthyomass culmination. In bream populations, a similar phenomenon is observed, but not in all reservoirs. In those in which fishing is not intensive, the periodicity in the onset of the age of ichthyomass culmination was not observed in bream. This can be clearly seen from the example of Ivankov and Uglich reservoirs:

Year	Age of ichthyomass culmination		Year	Age of ichthyomass culmination	
	Ivankov Reservoir	Uglich Reservoir		Ivankov Reservoir	Uglich Reservoir
1976	4	—	1982	6	6
1977	4	6	1983	6	6
1978	3	6	1984	3	4
1979	6	6	1985	5	5
1980	5	7	1986	4	5
1981	5	6	1987	6	7

In both reservoirs, over a period of 11–12 years, there was not a single case when some generation retained its leading position in respect of ichthyomass for 3 years or more in a row. Only once in Ivanko and twice in Ulgich reservoirs (marked by broken underline), did we observe generations retaining the position of the culminating group for two years. Moreover the magnitude of harvest of individual generation changed in a considerable range. For example, the number of three-year-old individuals, with which begins the age series analyzed by us, in Ivankov Reservoir varied in the range of 0.6–20.9 million, and in Uglich 0.1–3.3 million.

In the intensively exploited populations of bream, the periodicity of the age of ichthyomass culmination is observed fairly distinctly. In this respect, most demonstrative is the material given on next page on Volgograd and Tsimlyansk reservoirs.

In bream populations of Volgograd Reservoir, individual generations retained the leading position in respect of ichthyomass for four years in a row: 1969–1972 (shown with single underline) and three years: 1974–1976 (shown with double underline). These two periods are the longest. Besides these, there are three cases (shown by broken underline) where one generation retained leading position for two years. On the whole, in 13 out of 19 years, the age of ichthyomass culmination changed with a periodicity of 2–4 years. In bream of Tsimlyansk Reservoir, in view of a more prolonged period of observation, the periodicity is still more strongly manifest. One generation retained its culminating position of age classes for a period of five years: 1962–1966 (shown by

Year	Age of ichthyomass culmination		Year	Age of ichthyomass culmination	
	Volgograd Reservoir	Tsimlyansk Reservoir		Volgograd Reservoir	Tsimlyansk Reservoir
1962	-	7	1975	4+	5
1963	-	8	1976	5+	6
1964	-	9	1977	7+	7
1965	-	10	1978	7+	4
1966	-	11	1979	4+	5
1967	-	4	1980	5+	4
1968	-	5	1981	5+	5
1969	2+	6	1982	4+	5
1970	3+	7	1983	3+	6+
1971	4+	4	1984	3+	7+
1972	5+	5	1985	4+	6+
1973	4+	4	1986	3+	3+
1974	3+	4	1987	4+	4+

single underline); two generations, for four years: 1967–1970 and 1974–1977 (shown by double underline); and one generation, for three years: 1982–1984 (shown by three underlines). In addition four generations retained their leading position for two years (shown by broken underline). Out of 26 years of observations in 24 years bream population of Tsimlyansk Reservoir retained periodic change of the age of ichthyomass culmination for a period of 2–5 years.

However, it is not appropriate to attribute the phenomenon of the observed periodicity only to the effect of intensive fishing. For example, in Lake Il′men′, under intensive fishing, the periodicity of change of the age of ichthyomass culmination in bream population is not observed. Hence, during analysis of the process of change of the index under question, in order to establish the reasons for change of the noted periodicity (or its absence) it is necessary to take into account the ecological factors determining the level of harvest of generations and the pressure of elimination of age groups together with evaluation of the possible effect of fishing.

A population is an aggregate of individual generations with a definite number of individuals and biological indices (rate of weight gain, rate of elimination, etc.) typical of each of them. Hence a smooth (leveled) change of number of individuals and ichthyomass with age is possible only, as demonstrated by the data presented in Table 28, by averaging the data over many years. While examining any single year's data, the structure of population appears much more complex. This remark is much more valid for those species of fish, which, like bream, have a long life cycle. In the structure of such populations it is possible to identify many age classes with higher ichthyomass. One of them having the largest ichthyomass is a culminating group while others are subculminating. Each such age class corresponds to a definite generation differing in the level of harvest from those contiguous to it.

It is necessary to note that the presence of a group of culminating and subculminating age classes in respect of ichthyomass in an annual section of the structure of population, generally speaking, is not obligatory. In fish with long life cycles, there are years when the number of individuals and ichthyomass change with age more smoothly, without any jumps, like the averaged multiyear data. Hence during analysis of the population structure for many years different versions are observed, which are simple as also complicated by additional subculminating age classes in respect of ichthyomass.

In Table 29 we present the examples of different versions of the population structure of bream: without subculminating age class in respect of ichthyomass, with one and two older, and with two younger and two older subculminating age classes. Other combinations are also observed for subculminating and culminating age classes in respect of ichthyomass.

Unfortunately, in fishing-related study of bream populations attention is not focussed on the above peculiarities of its population structure. In essence, while developing forecasts of possible catches and determination of limits of permissible catches, it is imperative to take into account the annual changes in the population structure and recommend the optimal fishing regime with due weightage to these changes.

Mass maturation in bream populations sets in the age of 6–6+ to 10–10+. For this index, it is characteristic to experience geographic variation. The earliest mass maturation is observed in bream of Tsimlyansk Reservoir (6–6+) and the latest in Lake Il'men' (10–10+). In populations from other reservoirs, this index increases from south to north. An inverse relationship is observed for the number of possible age classes older than the time of mass maturation. From average multiyear data (Table 28), the age series of bream from Lake Il'men' comprises five mature age classes (including the one which attained maturity). In Ivankov, Uglich, and Gorkii reservoirs such classes number 5–6, in Cheboksar 8, Kuibyshev 9, Saratov and Volgograd 10 and in Tsimlyansk 13. In analyzing the data for each population in individual years, the maximum number of mature age classes in bream from Lake Il'men' was 9, and from Tsimlyansk Reservoir 17. In bream from other Volgograd reservoirs this index changed from 7 in Ivankov to 12 in Volgograd.

A similar geographic variation is also observed in the number of age classes in which ichthyomass culmination differs from the time of onset

Table 29. Number of individuals and ichthyomass of age groups in populations of bream from some reservoirs and Lake Il'men' (data of individual years)

Age	Lake Il'men', 1975		Lake Il'men', 1977		Tsimlyansk Reservoir, 1980		Kuibyshev Reservoir, 1981	
	No. in millions	Tons	No. in millions	Tons	No. in millions	Tons	No. in millions	Tons
2–2+	11.26	439	32.22	1,267	1.52	149	16.87	1,332
3–3+	4.75	418	3.38	297	9.75	1,832	1.33	150
4–4+	2.42	363	1.65	247	9.65	3,137	1.82	364
5–5+	2.27	496	1.32	289	2.96	1,745	8.09	2,133
6–6+	1.77	530	0.70	211	2.04	1,713	4.55	1,638
7–7+	4.41	1,667	1.46	551	2.00	2,047	8.19	4,004
8–8+	2.14	1,085	0.81	410	0.84	1,098	1.05	525
9–9+	0.90	588	0.47	304	0.72	1,330	2.10	1,270
10–10+	0.50	410	0.22	185	0.96	1,844	3.99	2,573
11–11+	0.06	67	0.03	34	0.26	585	1.19	880
12–12+	0.01	12	0.01	8	0.22	504	0.63	504
13–13+	0.01	5	0.01	6	0.04	104	0.14	116
14–14+	—	—	—	—	0.02	49	0.07	62
15–15+	—	—	—	—	—	—	0.28	235
16–16+	—	—	—	—	—	—	0.07	88
17–17+	—	—	—	—	—	—	0.21	290

Note: In the column "tons", the culminating age class is marked with double and subculminating with single underline.

of mass maturation. From the average multiyear data in bream populations from Lake Il'men', mass maturation sets in 8 years after ichthyomass culmination while in Tsimlyansk Reservoir it sets in after one year. The observed correlation between the age of mass maturation of bream and other population-structure indices are presented in Table 30. In it I have included data on the maximum range of age series, which also changes from south to north.

Geographic variation of population-structure indices in bream are not taken into consideration in studies aimed at developing the optimal regimes of exploitation of stocks of this fish. Often, the conclusion arrived at, during a study of bream in reservoirs of one geographic zone, are extrapolated to all other zones. As will be seen from Table 30 and a discussion that follows, such an approach is not entirely correct. Problems of regulating fishing and completeness of utilization of the productive potentials of bream populations have to be solved not in general terms but as applicable to a specific large reservoir. A generalized approach is permissible for small lakes situated in a single geographic

Table 30. Geographic variation of some population-structure indices in bream

Water body	Maximum range of age series, years	Age of mass maturation	Number of mature age classes		Difference in the age of ichthyomass culmination and mass maturation (average)
			Range	Average	
Lake Il'men'	14+	10+	4–5	5	8
Ivankov Reservoir	15	9	3–7	6	5
Uglich Reservoir	15	9	3–7	5	3
Gorkii Reservoir	16	8	5–9	6	4
Cheboksar Reservoir	16	8	6–9	8	5
Kuibyshev Reservoir	17+	9+	4–9	9	3
Saratov Reservoir	19	9	5–11	10	3
Volgograd Reservoir	18+	7+	5–12	10	3
Tsimlyansk Reservoir	22	6–6+	6–17	13	1

Note: In the last two columns, the average relates to multiyear data presented in Table 28.

zone. However, even in this case, it is necessary to aim at dividing these small lakes into individual fishery or limnological type and fishery measures should be linked with peculiarities of each of these lakes.

One of the indices, which is of great importance during analysis of the production processes in fish populations, is the ratio of ichthyomass of immature and mature age classes. In the former, we include those classes which are younger than the age of onset of mass maturation; the latter includes age classes starting from the age of mass maturation and older. In such a division, some part of the mature fish is likely to be included under immature class and vice versa. I use such a division because it provides greater simplicity of conducting estimates and does not add considerable distortions to the final conclusion. Ichthyomass of immature age classes was determined starting from 3–3+ so as to obtain comparable data in all bream populations. However, in this case, the total ichthyomass of the immature part of populations decreases due to fish of the age 0–0+, 1–1+ and 2–2+ being excluded from calculations. This fact must be borne in mind while discussing the data presented below.

Ichthyomass of the immature part of bream populations from Lake Il'men' and Volga reservoirs, from the average multiyear data, is 2.5–9.8 times higher than that of the mature part and only in Tsimlyansk Reservoir it is less. The maximum domination of immature fish is noticed in Lake Il'men' and Ivankov Reservoir.

On examination of the ichthyomass of both groups of fish from the average multiyear data for all the nine reservoirs the same geographic variation was identified as was noticed for other population-structure indices. For brevity I give below figure characterizing the ratio of ichthyomass of immature and mature fishes for individual populations in the same sequence as listed in Table 30: 9.0; 9.8; 3.7; 4.4; 3.8; 2.6; 3.3; 2.5; 0.9. It must be emphasized that the values are lower because of the incomplete consideration of the ichthyomass of the youngest age classes. However, the general tendency of variation, depending on this, remains unchanged.

A parallel may be drawn between the geographic variation of the ratio of these two parts of the population and the number of mature age classes. From south to north, the number of mature age classes decreases in the bream population, because of which their total ichthyomass also decreases. Parallel to this, the number of immature age classes increases as also their total ichthyomass. The ratio of ichthyomass of immature and mature fish increases in full conformity with both the processes during a transition from reservoirs in the south to those in the north.

The multiyear average data shows that ichthyomass culmination precedes the age of onset of mass maturation in all the bream populations (Table 28), because of which their structure corresponds to the third type. With transition from south to north, the difference in the ages of ichthyomass culmination and mass maturation increases and thereby the degree of correspondence of the structure to the third type seemingly rises.

The question arises: how constant is such a state of the structure when examining annual "sections" of bream populations in each reservoirs It seems that in all the reservoirs, except Kuibyshev, Volgograd and Tsimlyansk, in all years of observation, ichthyomass culmination precedes mass maturation. In Volgograd Reservoir in 2 out of 19 years of observations these two events did not coincide in time. In bream populations of Kuibyshev Reservoir in 14 out of 16 years of observation ichthyomass culmination preceded the onset of mass maturation; in one year it succeeded the latter and in another year both events coincided in time. Both coincidence and increase of the age of mass maturation was observed only in relation to one high-harvest generation of 1974, which, thanks to its vast number of individuals, occupied the leading position also in mass maturation a year later.

The most complex picture in relation to the correlation of the age of ichthyomass culmination and mass maturation is observed in bream populations from Tsimlyansk Reservoir. In 14 out of 26 years of observations, ichthyomass culmination preceded the age of mass maturation;

in 4 years it coincided, and in 8 years succeeded it. This peculiarity is due, firstly, to an early maturation of bream in Tsimlyansk Reservoir as compared to the northern reservoirs and, secondly, to the level of harvest of individual generations. A definite link is also observed with the above noted periodicity of changes of the age of ichthyomass culmination. All the eight cases of increase of the time limit for mass maturation fall in these periods of changes of the age of ichthyomass culmination, that extend for 3–5 years. All these cases are linked to the peculiarities of age-related dynamics of the ichthyomass of the high-harvest bream generation. Most demonstrative in this regard is the period 1962–1966 (see page 81), which accounts for 5 out of 7 years of higher age of mass maturation. This period is linked with one of the high-harvest generations of bream, which appeared in the initial years of the establishment of Tsimlyansk Reservoir. If we do not consider these five years, for which an increase of the age of mass maturation of the group of fish leading in ichthyomass is attributed to special circumstances, throughout the remaining period of observations (21 years) in bream populations of Tsimlyansk Reservoir, the age of ichthyomass culmination in rare cases (three years) succeeded the period of onset of mass maturation. All these cases are also linked with the individual high-harvest generations.

Thus, there is every basis to consider that the structure of bream populations from Kuibyshev, Volgograd, and Tsimlyansk reservoirs basically corresponds to the third type as in other Volga reservoirs and Lake Il'men'. In rare cases, upon appearance of individual high-harvest generations, does it change over to the second type.

ROACH

Roach is one of the most numerous fish of Volga reservoirs. However its reserves have been throughout under-utilized [Kuderskii and Yankovskaya, 1989]. In fishery statistics it is not separately considered but included in a group of small ordinary fish comprising other small and slow growing fish, in particular white bream and blue bream. Often one comes across a situation when the large roach is considered separately, and the small in a composite group of small ordinary fish. The importance of small ordinary fish, and more so of roach, in the overall catches of fish can be seen from the average annual data of 1981–1985 presented on next page.

The data on catches do not reflect the actual place of roach in the reservoir ecosystems, since the level of exploitation of its reserves is below the possible limit. Hence, the effect of fishing on the structure of roach populations, if it is there, is limited in nature. Inadequate attention to roach as an object of fishery also affects the extent of its study. Accumulation of information on various aspects of its biology is more limited than for leading commercial fish. In view of this, below I present data on the structure of roach populations only from five Volga reservoirs.

Reservoir	Average annual catch of small ordinary fish, ton	% of all fish as total average annual catch
Ivankov	72.7	29.3
Uglich	157.4	62.9
Gorkii	212.8	49.7
Cheboksar	142.5	75.5
Volgograd	1181.9	35.8

The number of age classes in the investigated population of roach, based on the multiyear average, varies insignificantly and constitutes from 10–10+ to 14–14+. An examination of the population structure of this species in individual reservoirs year by year reveals that the range of variation of the age series increases to 7+–14+. The limiting age of roach observed in these reservoirs practically agrees with the one established for its populations from the small lakes. The correlation between the range of the age series and the extent of commercial exploitation of reserves is not observed, which, not in the least, is due to the weaker roach catches in all the reservoirs under examination.

In roach populations, as in other species of fish, the rate of decrease of the number of individuals with age is higher than that of ichthyomass. The latter can be clearly visualized from the example of populations of this species from Volgograd Reservoir (average multiyear data).

Age groups of fish	1–1+ 2–2+	2–2+ 3–3+	3–3+ 4–4+	4–4+ 5–5+	5–5+ 6–6+	6–6+ 7–7+	7–7+ 8–8+	8–8+ 9–9+	9–9+ 10–10+
Ratio of number of individuals	5.5	3.7	1.0	1.1	1.9	2.6	1.9	4.3	4.0
Ratio of ichthyomass	0.8	2.0	0.7	1.1	1.4	1.6	1.3	4.5	2.9

Because of the above noted differences in roach populations from Ivankov, Gorkii, Cheboksar and Volgograd reservoirs, by the age of mass maturation the number of fish decreases to 10/17–10/149 and ichthyomass decreases from 0–10/33. Roach is an unprotected species of fish whose fishing is not regulated by any restrictions whatsoever. Hence arriving at any practical conclusions from these correlations is not necessary. But, all the same, for purpose of comparison, mention must be made of high losses of ichthyomass by the time of mass maturation.

The age of ichthyomass culmination in populations of roach from reservoirs, from the average multiyear data, is somewhat higher than that from small lakes, and in three cases it is 2–2+ (Table 31). In roach from Ivankov and Uglich reservoirs, it rises up to 3–3+ and 6–6+, respectively. One such deviation from the value, typical for other populations, is partly due to the inadequately complete assessment of the younger age groups. This conclusion is supported by the data on the change of number of young ones of roach in Ivankov and Uglich reservoirs, on one hand, and the remaining, on the other (Table 31).

Among the causes leading to an increase in the age of ichthyomass culmination in roach populations from reservoirs, one of the important causes is the higher weight gain. In small lakes, the food resources of roach are limited; it, as a rule, experiences food shortage and its weight gain is much lower than what is potentially possible. In conditions of reservoirs, different age groups of roach are provided with food better, starting from the stage of plankton feeding. In view of this, the rate of growth of mass increases considerably and the average weights in individual age classes increase (Table 32). For example, in Uglich Reservoir, the average mass of three-year-olds increases almost three fold and of four-year-olds four fold in comparison with roach from Lake Pelyuga. The increase of average mass of fish directly leads to an increase of ichthyomass of various age classes, which, in turn, causes a forward shift of the age of ichthyomass culmination. Hence in roach populations with higher rates of weight gain, ichthyomass culmination is observed in more older age classes.

In roach populations from reservoirs, the age of ichthyomass culmination is also affected by the degree of elimination of young ones, which is different than in small lakes.

The age of ichthyomass culmination in roach populations, in individual years of observation, varies in wide limits. In Ivankov Reservoir, it changes from 3+ to 7+, in Uglich in the range of 5+ –7+ and Volgograd 1+ –6+. Only in Gorkii and Cheboksar reservoirs, in years of observation, does it remain constant at two full years. How frequent is ichthyomass culmination in individual age classes can be seen from the data presented in Table 33. Analysis of these data also allows us to conclude that the age of ichthyomass culmination in roach populations changes according to geographic zonation. In Volgograd Reservoir, it is, on the average, less. Moreover, in roach populations of this reservoir, the minimum age of ichthyomass culmination drops to 1+ and 2+, which is not observed in Ivankov and Uglich reservoirs. The maximum value of this index also decreases. Unfortunately, the absence of data on other

Table 31. Number of individuals and ichthyomass of age groups in roach populations from reservoirs

Age	Ivankov Reservoir, 1973–1987		Uglich Reservoir, 1976–1982		Gorkii Reservoir, 1981–1987		Cheboksar Reservoir, 1986–1987		Volgograd Reservoir, 1969–1987	
	Million	Ton	Million	Ton	Million	Ton	Million	Ton	Million	Ton
1–1+	3.89	36.3	0.33	4.2	64.15	236.8	44.49	355.9	72.40	413.6
2–2+	4.20	71.5	1.49	21.4	44.72	1025.9	25.20	756.0	13.06	476.9
3–3+	5.28	174.1	2.21	81.1	16.62	704.6	11.15	446.0	3.52	305.4
4–4+	3.17	156.9	2.80	161.3	4.30	307.0	3.45	345.0	3.61	382.1
5–5+	1.49	115.1	2.63	217.6	1.31	168.3	0.98	157.4	3.15	350.5
6–6+	0.68	81.1	2.22	268.9	0.63	109.1	0.59	119.4	1.64	245.8
7–7+	0.42	76.7	0.65	113.4	0.35	76.1	0.21	56.8	0.64	105.8
8–8+	0.15	50.9	0.40	90.9	0.19	51.5	0.10	29.5	0.34	116.7
9–9+	0.04	13.6	0.12	39.9	0.09	29.6	0.07	27.2	0.08	25.3
10–10+	0.03	5.9	0.09	39.0	0.04	14.8	0.02	13.6	0.02	8.8
11–11+	0.005	2.2	0.01	62.3	0.02	7.3	0.01	6.8	—	—
12–12+	—	—	—	—	0.01	3.6	0.003	0.7	—	—
13–13+	—	—	—	—	0.002	1.2	0.001	0.7	—	—
14–14+	—	—	—	—	—	—	+	0.3	—	—

Note: Under the column "million" a single underline denotes age groups that reached mass maturity, under the column "ton" double underline denotes ichthyomass culminating group of fish.

Table 32. Weight gain of roach in some reservoirs and small lake Pelyuga, g

Age	Lake Pelyuga, 1984	Ivankov Reservoir, 1976–1977	Uglich Reservoir, 1976	Gorkii Reservoir	Volgograd Reservoir, 1972–1974
1–1+	6.3	19.9	12.5	6	—
2–2+	12.2	20.2	34.6	21	50
3–3+	22.7	29.0	48.1	44	68
4–4+	37.6	36.6	57.2	75	88
5–5+	58.6	45.5	76.3	112	128
6–6+	88.2	54.3	130.6	157	166
7–7+	111.7	103.0	—	208	256
8–8+	—	129.0	455	265	301
9–9+	—	—	—	329	394
10–10+	—	—	—	399	409
11–11+	—	—	—	475	430

Note: For this table, use was made of the data for Lake Pelyuga [Pechnikov, 1986], Ivankov and Uglich reservoirs [Baranova-Filon, 1980], Gorkii Reservoir [Bandura and Shibaev, 1984], and Volgograd Reservoir [Nebol'–sina, Elizarov and Abramova, 1980].

Table 33. Age of ichthyomass culmination (years) in roach population from reservoirs in years of observation

Reservoir	Age of ichthyomass culmination							Average age of ichthyomass culmination	Number of years of observation
	1+	2+	3+	4+	5+	6+	7+		
Ivankov	—	—	5	2	6	—	2	3.8	15
Uglich	—	—	—	2	3	2	—	5.0	7
Volgograd	4	3	3	3	6	—	—	3.2	19

reservoirs, including the ones in the south, does not permit discussion of the geographic variation of this and other population-structure indices in as much detail as was done in the case of bream.

In roach populations of Volgograd Reservoir, two maxima are noticed in the age of ichthyomass culmination. The first maximum occurs in the age class 1+ and the second in 6+. It may be suggested that, in this reservoir, this fish is represented by slow- and fast-growing groups of individuals. For each of these groups, there is a characteristically peculiar dynamics of the ichthyomass according to age classes.

The periodicity of change of the age of ichthyomass culmination in roach populations is expressed less distinctly than in bream. It is not observed in Ivankov, Gorkii and Cheboksar reservoirs. But in Uglich Reservoir, in three (1976, 1977 and 1978) out of seven years, one generation occupied the leading position, the ichthyomass culmination in these years changed correspondingly from 4+ to 5+ and 6+.

In roach populations from Volgograd Reservoir, the periodicity of change in the age of ichthyomass culmination is more clearly expressed:

Year	Age of ichthyomass culmination	Year	Age of ichthyomass culmination	Year	Age of ichthyomass culmination
1969	3	1976	2	1983	1
1970	4	1977	4	1984	5
1971	5	1978	3	1985	5
1972	3	1979	4	1986	4
1973	3	1980	5	1987	5
1974	4	1981	1		
1975	2	1982	2		

Over a period of 19 years, in Volgograd Reservoir, two generations were observed which led to ichthyomass for three years at a stretch, having formed the corresponding two periods (single and double underline). In three cases the duration of the period was two years (marked by broken underline).

Thus the effect of the level of harvest of generations of the age of ichthyomass culmination and appearance of periodicity of its change is observed not in all populations of roach.

Similarly, it was not possible to observe the level of harvest-dependent appearance of subculminating age groups in respect of ichthyomass in all populations of roach from reservoirs. This event was seldom observed in roach from Ivankov Reservoir but frequently in Volgograd Reservoir. In the latter reservoir, in roach populations, together with older it was sometimes possible to come across younger sub-culminating age classes. But on the whole, for roach the appearance of such culminating age classes is less characteristic than for bream.

In roach mass maturation in Ivankov, Uglich, Gorkii and Cheboksar reservoirs occurs in the age 4–4+ while in Volgograd Reservoir it is 3–3+. For sake of comparison, it may be pointed out that, in roach from Tsilmyansk Reservoir, the onset of mass maturation also coincides in time with the age 3–3+. Since the range of the age series in roach populations, based on the multiyear average data, changes in relatively narrow limits, even the number of age classes (starting with those having reached mass maturity), according to these figures, varies insignificantly: from eight in Ivankov, Uglich and Volgograd reservoirs to ten in Gorkii and 11 in Cheboksar reservoirs. When this index is examined in the annual section, the range of its variation increases. In roach populations from Ivankov Reservoir, the number of mature age classes in individual years varies in the range of 4–9 (often 7–9), in Uglich 5–9 and in Volgograd Reservoir it is 4–8 (often 6–8). As mentioned earlier, roach is a numerous fish everywhere. Hence it may be said that the change of number of mature age classes in the above limits does not affect the efficiency of reproduction of reserves of this fish.

The age of ichthyomass culmination, according to multiyear average data, in roach populations from four reservoirs precedes the mass maturation by one–two years (Table 31). In Uglich Reservoir, it succeeds mass maturation by just two years. In other words, in four reservoirs, the structure of roach populations, from the average data, belongs to the third type while in Uglich Reservoir it is of second type. However, when examining the correlation of the age of ichthyomass culmination and mass maturation in annual sections, the picture becomes much more complex. Only in Gorkii and Cheboksar reservoirs ichthyomass culmination in roach populations in the years of investigation persistently preceded mass maturation. In Ivankov Reservoir, only in five out of 15 years of observation, did ichthyomass culmination occur before the onset of mass maturation, in eight years it occurred after; both events occurred concurrently of two years. In Volgograd Reservoir, for 19 years of observations, these figures were respectively 7, 8 and 4 years. In Uglich Reservoir, for seven years of investigation, the age of ichthyomass culmination either succeeded (five years) the onset of mass maturation or coincided in time (two years). The higher number of years during which the age of ichthyomass culmination has higher than mass maturation of fish in Ivankov and Uglich reservoirs may be partly associated with the incomplete estimates of the young ones. However to consider the change of structure of roach population from these reservoirs, as entirely due to this fact, does not seem possible. Hence it must be considered that in conditions of large reservoirs in plains, unlike the investigated small lakes, the structure or roach populations may correspond to third type as also to the second. In one and the same reservoir, but in different years, the structure may change over from one type to the other as in some of the bream populations discussed earlier. Such dynamics of the correlation of the age of both events is primarily (other conditions remaining the same) due to the level of harvest of individual generations and differences in the degree of elimination of the young ones.

WHITE BREAM

In reservoirs of the Volga system, white bream is found in large numbers. However, in the extent of its study, it stands lower than bream and roach. Little attention to this species is because white bream, like the entire groups of small ordinary fish, is weakly exploited in commercial fishery. Its reserves are utilized inadequately. A somewhat better position in respect of exploitation of white bream is observed in Tsimlyansk Reservoir. In view of this, it may be considered that fishing does not exert appreciable effect on the structure of white bream populations in reservoirs.

Table 34. Number of individuals and ichthyomass of age groups in white bream populations from reservoirs

Age	Ivankov Reservoir, 1976–1987		Uglich Reservoir, 1976–1982		Volgograd Reservoir, 1969–1987		Tsimlyansk Reservoir, 1969–1987	
	No. in thousands	Ton	No. in thousands	Ton	No. in thousands	Ton	No. in thousands	Ton
2–2+	343.8	5.8	107.2	1.8	15.9	446.0	0.8	186.6
3–3+	<u>324.5</u>	8.5	<u>249.6</u>	8.8	<u>6.3</u>	354.7	<u>5.0</u>	585.4
4–4+	436.4	21.2	1032.5	<u>43.3</u>	5.2	467.4	10.8	<u>1663.1</u>
5–5+	325.8	<u>21.8</u>	631.9	34.5	4.4	515.3	7.1	1468.6
6–6+	196.4	20.0	276.3	27.1	3.9	<u>558.9</u>	4.1	914.2
7–7+	114.1	15.8	185.9	23.9	2.4	425.8	1.8	500.6
8–8+	74.5	14.0	227.2	41.9	2.0	449.5	6.6	198.1
9–9+	25.9	6.5	76.0	14.2	1.3	361.8	0.2	102.2
10–10+	15.8	4.7	39.3	8.6	0.6	202.6	0.07	32.6
11–11+	4.4	1.4	16.6	6.8	0.2	104.9	0.01	6.1
12–12+	2.0	0.7	6.6	2.6	0.3	115.9	0.02	1.5
13–13+	—	—	—	—	0.002	0.7	+	0.3

Note: See note to Table 31.

In the white bream populations investigated from four reservoirs (Table 34) the number of age classes, from the average of multiyear data, reaches 12–12+ and 13–13+. Moreover, in Ivankov Reservoir, in one case this fish was caught at the age of 14+. In individual years the range of age series in all white bream populations from reservoirs changed from 8+ to 14+. On the whole, for reservoirs the range of age series in white bream populations is more prolonged than in small lakes. But, all the same, in this respect white bream is much inferior to bream. In spite of the fact that in many aspects of biology white bream is closer to bream, in the life span it is much closer to roach.

The number of individuals of white bream decreases with age at a much higher rate than the decrease of its ichthyomass. This is clearly seen from the example of population from Volgograd Reservoir:

Age groups of fish	2–2+, 3–3+	3–3+, 4–4+	4–4+, 5–5+	5–5+, 6–6+	6–6+, 7–7+	7–7+, 8–8+	8–8+, 9–9+	9–9+, 10–10+	10–10+, 11–11+	11–11+, 12–12+
Ratio of number of individuals	2.5	1.2	1.2	1.1	1.6	1.2	1.5	2.2	3.0	0.7
Ratio of ichthyo-mass	1.3	0.8	0.9	0.9	1.3	0.9	1.2	1.8	1.9	0.9

In white bream populations, the age of ichthyomass culmination, from the average of multiyear data, constitutes 4+–6+. However the complete range of variation of this character, considering data of individual years, reaches from 2+ to 9+.

Although from the averaged multiyear data the geographic variation of the age of ichthyomass culmination is not noticed, it manifests most convincingly when one considers the material on each white bream population on a year by year basis (Table 35). In Ivankov and Uglich reservoirs, the maximum age of ichthyomass culmination is equal to 8+ and the corresponding average is 5.9 and 6.6 years respectively. In Volgograd Reservoir, this index reaches a value of 9+, but for a large number of years (14 or 73.7%), it does not exceed 6+, and the average of 5.3 years. In Tsimlyansk Reservoir, ichthyomass culmination is observed in age classes not older than 6+, and the average age drops to 4.6 years.

The periodicity of change of the age of ichthyomass culmination in individual white bream populations is expressed to different extent and yet it is found in each one of them. As example, the data from Ivankov and Tsimlyansk reservoirs are presented below:

Year	Age of ichthyomass culmination Ivankov Reservoir	Tsimlyansk Reservoir	Year	Age of ichthyomass culmination Ivankov Reservoir	Tsimlyansk Reservoir
1969	-	4+	1979	4+	5+
1970	-	3+	1980	5+	6+
1971	-	4+	1981	6+	6+
1972	-	4+	1982	8+	5+
1973	-	5+	1983	6+	4+
1974	-	4+	1984	7+	3+
1975	-	5+	1985	4+	4+
1976	5+	6+	1986	5+	5+
1977	5+	5+	1987	6+	6+
1978	8+	4+			

Table 35, Age of ichthyomass culmination in white bream populations from reservoirs for years of observation (number of years)

Reservoir	Age of ichthyomass culmination								Average age of ichthyomass culmination	Number of years of observation
	2+	3+	4+	5+	6+	7+	8+	9+		
Ivankov	—	—	2	4	3	1	2	—	5.9	12
Uglich	—	—	1	1	1	1	3	—	6.6	7
Volgograd	3	—	3	4	4	3	1	1	5.3	19
Tsimlyansk	—	2	7	6	4	—	—	—	4.6	19

In Uglich Reservoir, despite short period of observation, one three-year period (1976, 1977 and 1978) of change of age of ichthyomass culmination and one two-year period (1980, 1981) were noticed. The periodicity of question is most markedly expressed in white bream populations of Tsimlyansk Reservoir. Here, during a period of 14 out of 19 years, two two-year periods (broken underline), two three-year periods (single and double underlines) and one four-year period (three underlines) were observed. More rare was the periodicity of change of the age of ichthyomass culmination observed in white bream populations of Volgograd Reservoir: for a period of 19 years there were only three two-year periods.

As mentioned earlier, the periodicity of change of the age of ichthyomass culmination is due to the generation leading in each reservoir, for many years in a row, in respect of the level of its harvest. Appearance of such generations should not be very frequent. With frequent appearance of leading generations, the periodicity in question disappears. Hence the absence, or rare occurrence of periodicity in the change of the age of ichthyomass culmination, in a multiyear observation of the structure of white bream populations, could serve as an index of the relative stability of reproduction of its reserves. On the other hand, a constancy in the appearance of such periodicity confirms the cyclic changes in the conditions of multiplication of white bream, development of its spawn and larvae, and the extent of elimination of its young ones.

Ichthyomass subculminating age classes in white bream populations are observed rarely. In most reservoirs, as a rule, they are not observed. In some years, old as well as young ichthyomass subculminating age classes were observed in Volgograd Reservoir. Populations of white bream differ from those of bream but are closer to roach in respect of the rare presence of subculminating groups.

A number of mature age classes (starting from the age of mass maturation of fish) in white bream populations, according to the average multiyear data, equals 10–11. In individual years their number was 7–12 in the populations from Ivankov, 6–10 in Uglich, 7–11 in Volgagrad and 6–11 in Tsimlyansk reservoirs. It changed 1.6–1.8 times. In the investigated white bream populations from small lakes, such age classes were 3–6 with an overall range of age series to 7–10 years. A comparison of the figures presented reveals that the total life span as also the attendant number of mature age classes in white bream populations are obviously associated with the type of the ecosystem of which this species is a part. However, these indices are not dependent on the population diversity. Thus, in Ivankov Reservoir, the ichthyomass of white bream (starting from the age class 2–2+), according to the average multiyear data, is 3.7, in Uglich it is 8.6, in Volgograd 11.5 and in Tsimlyansk 21.0 kg/ha. The differences between the extreme values are 5.7 times. However, the number of mature age classes as also the limiting life span show closer values.

According to the multiyear average data, culmination of ichthyomass in white bream populations of all reservoirs occurs one–three years after the onset of mass maturation. In view of this, they belong to the second type. They differ significantly in the above index from populations from small lakes, the structure of which corresponds to the third type.

The average multiyear data do not reflect the entire range of variation of the correlation of the age of ichthyomass culmination and mass maturation of white bream. Analysis of data for individual years rarely reveals deviation from the described multiyear average scheme. Thus, in white bream populations from Tsimlyansk Reservoir, in 2 out of 19 years of observation, these events coincided in time while in Volgograd Reservoir in 3 out of 19 years ichthyomass culmination preceded the age of mass maturation.

These rare deviations do not change the overall conclusion about the inclusion of white bream populations from reservoirs into the second type. However, considering the data from small lakes, it may be concluded that in white bream, depending on the habitat conditions, the structure of population may change from the third type to the second (and vice versa).

BLUE BREAM

The population structure of blue bream was studied on the material from Tsimlyansk Reservoir. The material was collected during 1968–1987. In this reservoir, blue bream has great fishery value and its reserves are generally utilized intensively [Fesenko, 1976]. The average yearly catch during 1981–1985 was 795 ton or 6.2% of the total catch of fish.

In blue bream population the number of age classes varies from year to year in the range of 8–13. In nine out of 20 years of observation, the age series terminated with 12–13-year olds. In other words, the observed limiting age is not an exception. The variable range of age series in blue bream population is not only due to the random collection of material, which is usually relied upon in such cases, but also the unequal degree of elimination of fish of different generations.

The rate of decrease in the number of individuals in blue bream population is higher than that of ichthyomass:

Age groups of fish	1–1+, 2–2+	2–2+ 3–3+	3–3+ 4–4+	4–4+ 5–5+	5–5+ 6–6+	6–6+ 7–7+	7–7+ 8–8+	8–8+ 9–9+	9–9+ 10–10+
Ratio of number of individuals	0.4	0.7	1.0	1.8	1.1	2.0	2.3	2.8	4.0
Ratio of ichthyomass	0.2	0.4	0.7	1.4	1.0	1.7	2.1	2.1	4.5

By the time of onset of mass maturation, the number of individuals decreases to 5/9 the maximum while ichthyomass deceases to 5/7.

The age of ichthyomass culmination, according to the averaged data of 20 years, in blue bream population equals 4–4+; but in individual years it may change from 3–3+ to 7–7+. For the entire duration of observation, in 10 years ichthyomass culmination occurred in the age classes 3–3+ and 4–4+; in nine years in 5–5+, and 6–6+; and only in one year in 7–7+. In other words, despite considerable prolongation the age of ichthyomass culmination in half the cases sets in more younger age classes.

In blue bream population the periodicity of change in the age of ichthyomass culmination is expressed quite markedly in view of the appearance of individual generations leading in the level of harvest:

Year	Age of ichthyomass culmination	Year	Age of ichthyomass culmination
1968	6	1978	4
1969	6	1979	5
1970	3	1980	6
1971	4	1981	4
1972	4	1982	5
1973	3	1983	3+
1974	4	1984	5+
1975	5	1985	6+
1976	6	1986	3+
1977	7+	1987	4+

For 20 years of observations one five-year period (single underline), one three-year (double underline) and four two-year periods (broken underline) were identified. Their total duration works out to 16 years. This periodicity confirms the instability of reproduction of the blue bream reserves, which is also supported by the following data on the ichthyomass of two-year-olds (from 1983 three-year-olds):

Year	Ichthyomass, ton	Year	Ichthyomass, ton
1968	68.6	1978	58.3
1969	1,158.3	1979	335.0
1970	514.1	1980	96.3
1971	34.0	1981	334.8
1972	1,195.8	1982	89.0
1973	76.0	1983	261.8
1974	18.7	1984	55.0
1975	8.6	1985	731.3
1976	558.9	1986	9.3
1977	11.2	1987	100.5

Only in two years did the ichthyomass surpass 1000 tone and in three years over 500 ton, while for 12 years it remained at the level of 8.6–100.5 ton. The rare appearance of numerous populations results in leading position of individual generations for three and five years. Here, not only the absolute value of ichthyomass but, what is equally important, its ratio with the ichthyomass of neighboring generations also play a great role. Actually the 1974 generation contributed to ichthyomass of two-year-olds in 1976, of the order of 558.9 ton. The absence of other generations with comparable ichthyomass values placed the 1974 generation in leading position in respect of ichthyomass in the year 1978, 1979 and 1980. Similarly, the 1970 generation had in 1972 ichthyomass of 1195.8 ton in the age of two complete years. None of the subsequent generations were comparable to it in harvest. Hence the 1970 generation occupied the leading position for five years, i.e. 1973–1977. At the same time, the 1968 generation, which in 1970 gave an ichthyomass of 514.1 ton for the two-year-olds, did not reach the leading level since this was "obstructed" by the high-harvest 1970 generation.

The structure of population of blue bream in individual years becomes complex on account of the appearance of subculminating age groups in respect of ichthyomass. Some of the variations arising thereby are presented in Table 36. In 1984, as also in the 20 year average, the structure of blue bream population was simple: only one ichthyomass culminating age class was present. In other years, besides it there were older and younger (or only older) subculminating age groups. Such a

Table 36. Number of individuals and ichthyomass of age groups in blue bream population of Tsimlyansk Reservoir

Age	1971		1980		1984		Average for 1968–1987	
	No. in million	Ton	No. in million	Ton	No. in million	Ton	No. in million	Ton
1–1+	2.48	61.9	0.51	12.7	0.31	11.6	0.95	41.4
2–2+	0.34	34.0	1.01	96.3	0.47	55.0	2.73	285.8
3–3+	6.91	1,416.6	7.44	1,152.6	0.64	119.7	3.97	734.0
4–4+	8.89	2,622.0	0.51	112.6	1.22	305.1	4.03	990.8
5–5+	0.25	96.3	0.34	94.6	2.80	883.5	2.22	734.4
6–6+	0.43	198.7	6.08	2,068.6	0.87	320.7	2.02	743.8
7–7+	1.71	920.7	0.07	27.2	0.90	369.7	1.03	451.8
8–8+	1.66	1,012.0	0.27	125.7	0.12	55.7	0.44	220.1
9–9+	0.02	16.0	0.51	266.2	0.16	76.6	0.16	107.0
10–10+	0.01	8.6	0.17	97.7	0.05	26.4	0.04	23.7
11–11+	0.01	9.1	—	—	0.002	1.3	0.01	7.6
12–12+	0.01	9.8	—	—	0.02	9.6	0.02	17.9
13–13+	0.01	10.3	—	—	0.004	3.0	0.01	10.1

Note: See note to Table 29.

complex structure of blue bream population is linked with the level of harvest of individual generations.

Mass maturation in blue bream population is observed in the age class 5–5+. Since the range of the age series reaches 13–13+, the number of mature age classes (starting from the age of mass maturation), from the average data for 20 years, was nine (Table 36). In individual years, this number may vary from 4 to 9; for ten years it was 7–9, and for other 10 years from 4 to 6.

According to the multiyear average data, ichthyomass culmination in blue bream population precedes mass maturation by one year (Table 36). However, in individual years the time correlation of these two events is different. For 20 years of observation, ichthyomass culmination preceded mass maturation in ten years, in six years it succeeded, and in four years both events occurred concurrently. In other words, the structure of population of blue bream in the multiyear section does not remain stable but oscillates between the third and second types. In view of this, the ratio of ichthyomass of immature and mature parts of the population also becomes unstable from year to year. Although, on the average for 20 years, the ichthyomass of immature fish was slightly lower than that of mature fish (ratio was 0.90) but in individual years this ratio shows considerable variation.

The ichthyomass of immature part of the population was less for ten years. Its ratio to ichthyomass of the mature part was 0.03–0.73, average being 0.30. In the remaining ten years it predominated. Its ratio to ichthyomass of mature part of the population was 1.09–9.73, average being 1.87. Despite the incomplete estimate of younger age groups during determination of the number of blue bream as well as exclusion of age class 0+ from the calculation, the presented figures reflect a general picture of the ratio of ichthyomass of both parts of blue bream population.

As will be seen from Table 37, the ichthyomass of the mature part of the population of blue bream does not depend on the number of mature age classes. It is determined by the correlation of the ichthyomass culmination and mass maturation. In years when the age of ichthyomass culmination is higher than the age of onset of mass maturation or coincides with it, the mature fish predominate in ichthyomass. The structure of population in these cases corresponds to the second type. In

Table 37. Yearly change in the correlation of ichthyomass of immature and mature parts of blue bream population

Year	Number of mature age classes	Correlation of age of ichthyomass culmination and mass maturation	Ratio of ichthyomass of immature to mature fish
1968	9	Higher	0.48
1969	9	Higher	0.71
1970	9	Lower	1.56
1971	9	Lower	1.81
1972	9	Lower	1.74
1973	5	Lower	5.27
1974	5	Lower	9.73
1975	6	Concurrent	0.03
1976	5	Higher	0.17
1977	4	Higher	0.48
1978	6	Lower	1.85
1979	5	Concurrent	0.12
1980	6	Higher	0.51
1981	6	Lower	1.28
1982	8	Concurrent	0.61
1983	6	Lower	1.33
1984	9	Concurrent	0.13
1985	7	Higher	0.73
1986	8	Lower	1.09
1987	8	Lower	1.19

years when the age of ichthyomass culmination is lower than that of mass maturation, immature fish predominate in ichthyomass. The structure of such population changes to the third type.

Thus, unlike the white bream populations from reservoirs whose structure basically corresponds to the second type, the only in individual years to the third type, the structure of blue bream population seems to be quite labile. In one and the same reservoir, in different years, it changes, with equal frequency, from the third to second type. In this respect, blue bream from Tsimlyansk Reservoir shows similarity with roach from Volgograd Reservoir.

PIKE

The structure of pike populations has been examined from the material collected from five reservoirs. The number of individuals and ichthyomass of pike is not very high in them, and is of secondary importance to fishery. In the average annual catch for the period 1981–1985, the share of pike was only 0.5–6.8%:

Reservoir	Average annual catch of pike, ton	% of average annual catch of fish
Ivankov	1.2	0.5
Uglich	4.1	1.6
Gorkii	15.2	3.6
Cheboksar	12.8	6.8
Kuibyshev	180.9	4.0

Pike belongs to the group of fish with very long life cycle. However, in conditions of the above mentioned five reservoirs, the limiting age did not exceed 12+. In Kuibyshev Reservoir, the maximum age of pike was 8+ while in Ivankov Reservoir it was even 6+. However, in Cheboksar Reservoir pike population showed up to 15 age classes [Zaloznykh, 1985a]. That is, in the range of the age series, pike from reservoirs is similar to that from small lakes.

The rate of decrease of the number of individuals and ichthyomass with age is presented below from the example of pike population from Cheboksar Reservoir:

Age groups	1–1+, 2–2+	2–2+ 3–3+	3–3+ 4–4+	4–4+ 5–5+	5–5+ 6–6+	6–6+ 7–7+	7–7+ 8–8+	8–8+ 9–9+
Ratio of number of individuals	2.8	2.8	1.9	2.2	2.7	2.2	2.1	2.4
Ratio of ichthyomass	0.4	1.3	1.1	1.4	1.5	1.6	1.5	1.8

Unlike small lakes, in pike from Cheboksar Reservoir, the rate of decrease of the number of individuals and ichthyomass are higher. In view of this, in populations from reservoirs instead of the gradual decrease of ichthyomass with age, which is characteristic of small lakes, a relatively faster decrease is observed (compare Tables 12 and 38). This noted difference is due to a higher elimination of older age groups of pike in reservoirs under the impact of fishing.

In pike populations from reservoirs, the age of ichthyomass culmination, according to the multiyear average data, varies from 2–2+ to 5–5+. In Kuibyshev Reservoir, in individual years, ichthyomass culmination was observed in the age class 6–6+. Considering the variation of this index in individual populations from year to year, the overall limits of its variation in reservoirs seem almost the same in small lakes.

In pike population of Kuibyshev Reservoir, a clearly expressed periodicity is observed in the change of age of ichthyomass culmination from year to year:

Year	Age of ichthyomass culmination	Year	Age of ichthyomass culmination
1979	4+	1984	5+
1980	5+	1985	5+
1981	6+	1986	5+
1982	3+	1987	6+
1983	4+		

Despite the short period (nine years) of investigation, in this population one clearly observes two three-year (single and double underline) and one two-year periods (broken underline) of change in the age of ichthyomass culmination. This periodicity is associated with the appearance of individual generations differing in their level of harvest from the neighboring ones.

In pike populations from reservoirs the mass maturation of females occurs in age classes from 2–2+ to 5–5+. These figures agree with the range of variation of the age of the onset of mass maturation in the fish species under investigation from small lakes. An identical picture is observed when one compares the correlation of the age of ichthyomass culmination and mass maturation in pike from reservoirs and small lakes. According to the average multiyear data, in pike population from three out of five reservoirs ichthyomass culmination precedes mass maturation; in one it succeeds and in another both these events occur concurrently.

In individual reservoirs, the correlation of the age of ichthyomass culmination and mass maturation of pike changes from year to year.

Table 38. Number of individuals and ichthyomass of age groups in pike populations from reservoirs

Age	Ivankov Reservoir, 1976–1982		Uglich Reservoir, 1976–1982		Gorkii Reservoir, 1981–1987		Cheboksar Reservoir, 1981–1987		Kuibyshev Reservoir, 1979–1987	
	Thousand	Ton	Thousand	Ton	Thousand	Ton	Thousand	Ton	Thousand	Ton
1–1+	28.3	3.0	25.3	2.9	127.5	3.6	1,288.8	34.3	—	—
2–2+	44.3	9.4	58.1	13.7	60.2	11.2	465.7	82.7	63.7	19.8
3–3+	28.7	16.3	48.5	26.0	29.2	17.9	167.6	64.6	103.6	98.3
4–4+	5.8	10.7	14.5	13.9	11.9	16.9	87.2	58.1	143.2	193.4
5–5+	2.1	6.2	4.2	6.8	6.1	12.2	39.3	41.6	106.3	251.3
6–6+	1.0	4.4	0.8	2.5	2.0	5.5	14.8	28.7	62.1	227.1
7–7+	—	—	0.5	1.8	1.0	3.6	6.6	18.2	21.1	93.2
8–8+	—	—	0.3	1.4	0.5	2.9	3.1	11.8	7.1	36.3
9–9+	—	—	0.3	1.7	0.3	1.7	1.3	6.5	—	—
10–10+	—	—	—	—	0.1	0.5	0.1	0.7	—	—
11–11+	—	—	—	—	0.02	0.2	0.04	0.4	—	—
12–12+	—	—	—	—	—	—	0.01	0.1	—	—

Note: In the column "thousand", the age of mass maturation is identified by a single underline, while the ichthyomass culminating age group is identified with double underline under the column "ton".

Thus in Cheboksar Reservoir, in two out of seven years, ichthyomass culmination occurred before the onset of mass maturation, in three years after it and two years both events occurred concurrently. In Kuibyshev Reservoir, the age of ichthyomass culmination in pike populations, in most cases (in eight out of nine years of observation), succeeded the period of mass maturation and in one year coincided with it.

Because of the inconsistency of the correlation of the ages of ichthyomass culmination and mass maturation in different reservoirs, as also the variation in this correlation from year to year, the structure of populations of pike, in some cases, corresponds to the second type while in others to the third type. If such a conclusion for pike populations from small lakes appears entirely natural in view of the specifics of the intraecosystem links typical for this group of reservoirs, then as applied to the reservoir populations it at the first glance, seems rather unusual. It would be more natural to consider that the structure of pike populations from large reservoirs in plains always corresponds to the second type. However, as factual data reveal, pike in large reservoirs always behaves as fish with a short life cycle. This is related to peculiarities of reproduction of its stocks in conditions of reservoirs in plains, specific behavior (solitary predator) as well as effect of fishing.

VOLGA ZANDER

The structure of populations of Volga zander was studied in Kuibyshev, Volgograd and Tsimlyansk reservoirs, where it has considerable fishery importance. To evaluate the role of Volga zander in overall catches of fish, the following data is presented:

Reservoir	Average annual catch of Volga zander, ton	% of average annual catch of fish
Kuibyshev	192	4.2
Volgograd	306	9.3
Tsimlyansk	918	7.1

The extent of exploitation of stocks of Volga zander is not the same in these reservoirs. It is exploited most intensively in Tsimlyansk Reservoir. At the same time, in Kuibyshev Reservoir the intensity of exploitation of this fish does not reach the optimum level. In view of this, the effect of fishing on the structure of its populations is not the same everywhere. The maximum effect may be expected in Tsimlyansk Reservoir and the least in Kuibyshev Reservoir.

The range of the age series in populations of Volga zander in individual years changes in Kuibyshev Reservoir from 6+ to 13+, in

Volgograd from 6+ to 8+, and in Tsimlyansk Reservoir from 7+ to 12+. At times, here one observes Volga zander with age up to 15+. Thus for Volga zander fishing at the existing intensity does not exert influence on the limiting number of age classes. In this respect, most demonstrative are the figures for Kuibyshev Reservoir. In it, up to the year 1970, the catch of Volga zander never exceeded 52 ton per year [Korablev, 1972]. In subsequent years its catch increased, in the period 1975–1985 it was annually 170.0 to 435.1 ton [Kuderskii and Khuzeeva, 1986]. However, in the first period, according to data of P.M. Chikova (1966), six–seven-year olds were found in the spawning stage while in 1982–1986 the population included 11–13 age classes of fish. In Tsimlyansk Reservoir, where stocks of Volga zander, for a long time, were exploited intensively [Lapitskii, 1970; Tyunyakov, Koval and Naumov, 1984], the maximum age of this fish at the end of 1960s was 14–15 years. During 1984–1987, it decreased to 9+–10+ but in 1983 rose to 15+.

The rate of decrease of the number of individuals and ichthyomass in Volga zander population is presented below from the average multiyear data from Tsimlyansk Reservoir:

Age groups	3-3+, 4-4+	4-4+, 5-5+	5-5+, 6-6+	6-6+, 7-7+	7-7+, 8-8+	8-8+, 9-9+	9-9+, 10-10+
Ratio of numbers	1.0	1.9	1.6	2.0	2.0	1.6	1.7
Ratio of ichthyomass	0.7	1.5	1.3	1.9	1.7	1.1	2.1

Age groups	10-10+, 11-11+	11-11+, 12-12+	12-12+, 13-13+	13-13+, 14-14+	14-14+, 15-15+
Ratio of numbers	1.9	1.6	1.5	1.2	5.0
Ratio of ichthyomass	1.6	1.5	1.5	1.1	4.1

As in other species of fish, the rate of decrease in the number of individuals in populations of Volga zander is higher than the rate of its ichthyomass.

According to the multiyear average data, ichthyomass culmination is observed in population of Volga zander from Kuibyshev in the age of 5–5+, Volgograd in 1–1+ and in Tsimlyansk reservoirs in 4–4+ (Table 39). In individual populations the age of ichthyomass culmination does not remain constant from year to year, and it changes in Kuibyshev Reservoir in the range of 4+–7+, in Volgograd 1+–5+ and in Tsimlyansk 3+–6+. The index in question is often 4+–5+ in Volga zander populations of Kuibyshev Reservoir (Fig. 10), the figures for Volgograd are 1+–3+ and for Tsimlyansk 3+–4+.

Table 39. Number of individuals and ichthyomass of age groups in
populations of Volga zander from reservoirs

Age	Kuibyshev Reservoir, 1971–1987		Volgograd Reservoir, 1971–1983		Tsmilyansk Reservoir, 1968–1987	
	No. in thousands	Ton	No. in thousands	Ton	No. in thousands	Ton
1–1+	—	—	5,792.3	946.9	—	—
2–2+	322.6	18.8	3,046.9	606.9	453.1	98.7
3–3+	485.3	73.1	1800.8	545.4	1,135.3	345.9
4–4+	1,130.5	320.4	691.5	444.6	1,132.9	468.3
5–5+	1,322.7	472.8	275.4	146.9	607.8	308.6
6–6+	631.4	292.3	277.9	175.4	380.0	238.7
7–7+	187.1	106.9	46.4	32.7	189.1	127.7
8–8+	52.3	48.0	18.7	6.9	92.4	73.3
9–9+	51.2	42.5	—	—	56.3	67.5
10–10+	14.5	15.7	—	—	34.0	32.5
11–11+	—	—	—	—	18.0	19.8
12–12+	—	—	—	—	11.6	13.3
13–13+	—	—	—	—	7.9	8.7
14–14+	—	—	—	—	6.5	8.2
14–14+	—	—	—	—	1.3	2.0

Note: See note to Table 38.

The periodicity in the change of the age of ichthyomass culmination is expressed very weakly in the populations of Volga zander. In Kuibyshev Reservoir, for 15 years of observation, the leading position of a generation in terms of ichthyomass was observed no more than for two years (three times). The same holds good in Volgograd Reservoir. Here for 13 years of observation, the Volga zander population showed two two-year periods. Only in Tsimlyansk Reservoir, for 19 years of observation there were two 3-year periods of change of ichthyomass culmination values (1973–1975 and 1981–1983) and five two-year periods.

The fact that a generation of fish retains its leading position for two years in a row does not characterize it as particularly significant leader according to the level of harvest, as compared to the neighboring generations and primarily with one directly following it. That it is really so, is seen from the example of Volga zander population from Tsimlyansk Reservoir. The highest-harvest generation of 1967 contributing in 1970 to the ichthyomass of three-year olds, equaling 1226.5 ton, exceeded ichthyomass of the similar age class of the directly following three generations by 1.6, 4.2 and 1.8 times respectively. The 1970 generation leading for three years had in 1973 an ichthyomass of 683.5 ton for

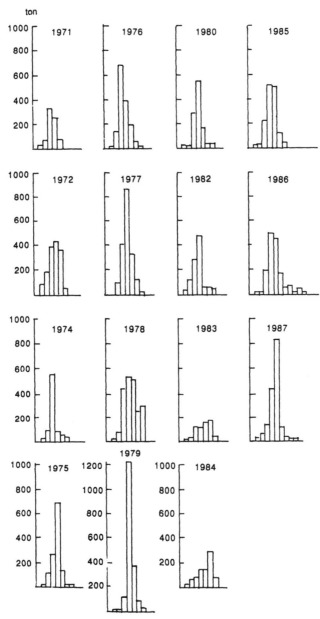

Fig. 10. Dynamics of ichthyomass of age groups in populations of Volga zander from Kuibyshev Reservoir.
The age class 2+ is shown in the extreme left.

three-year, olds and surpassed the ichthyomass of similar age class of the three succeeding generations by 4.1, 3.1 and 1.5 times respectively. As the example of blue bream shows, for a prolonged retention of the leading position, other correlations of the harvest level are necessary. Thus in bream from Tsimlyansk Reservoir, the two-year olds of the 1970 generation retained the leading position for five years with ichthyomass surpassing that of two-year olds of the succeeding three generations by 15.7, 64.0 and 130.1 times.

The ichthyomass subculminating age groups in populations of Volga zander are almost not developed. This simplifies their structure. It is because of the above described peculiarities in the harvest levels of individual (neighboring) generations.

In population of Volga zander from Kuibyshev Reservoir mass maturation was noticed in the age class 5–5+, in Volgograd at 3–3+, and in Tsimlyansk Reservoir at 4–4+. The number of mature age classes (beginning with the age of mass maturation) increases from north to south. The range of variation of this character in Volga zander from Kuibyshev Reservoir constitutes from two to eight age classes, in Volgograd it is four to five and in Tsimlyansk from four to 12. Their average number was 4.1, 4.9 and 9.5 respectively. Thus, the existing intensity of exploitation of Volga zander stocks in reservoirs does not affect the number of mature age classes in populations, like the overall range of the age series.

The correlation of the ages of ichthyomass culmination and mass maturation in populations of Volga zander, according to the average multiyear data, is quite peculiar. In two populations these events coincide in time. However, in Volgograd Reservoir ichthyomass culmination occurs earlier (Table 39). In view of this, the scales of variation of the correlation in question, in each of the three populations of Volga zander from year to year are of interest (Table 40). It is seen from Table 40, that in Volga zander from Kuibyshev and Tsimlyansk reservoirs, often (80.0 and 73.7% respectively) the age of ichthyomass culmination was more or less equal to the age of mass maturation. As against this, in Volga zander population from Volga Reservoir, in 7 (53.9%) out of 13 years of observation it was less.

The examined peculiarities allow us to consider that the population structure of Volga zander changes from the third type to the second and vice versa, depending on specific habitat conditions. Thus in Kuibyshev Reservoir, the structure of Volga zander population often corresponds to the second type [Kuderskii and Zhuzeeva, 1986]. The same conclusion holds true in relation to its population from Tsimlyansk Reservoir. On the other hand, in Volgograd Reservoir, the structure of population of this species usually belongs to the third type and rarely to the second type.

I think that one of the reasons for such inconsistency of the population structure is purely methodological in nature. An impression is

Table 40. Correlation of the ages of ichthyomass culmination and mass maturation in populations of Volga zander (number of years)

Reservoir	Correlation of age of ichthyomass culmination to mass maturation in generations			Number of years of observation
	Lower	Equal	Higher	
Kuibyshev	3	9	3	15
Volgograd	7	3	3	13
Tsimlyansk	5	7	7	19

created that the estimate of younger age groups of Volga zander (up to 3–3+ inclusively) in Kuibyshev and Tsimlyansk reservoirs was insufficiently complete. However, explaining structural inconsistency of Volga zander populations only through inadequacies of estimates of the number of individuals in younger age groups would not be entirely correct. There is no doubt that such inconsistency does not actually exist; but due to incomplete estimate of the scale it appears exaggerated. The basic reason of variation in the structure of Volga zander population lies in the peculiarities of the annually formed correlation of the ichthyomass of age classes of fish of different generations (primarily neighboring generations) and the pressure of elimination of the young ones.

PIKE-PERCH

In the conditions obtaining in reservoirs, pike-perch is one of the most valuable commercial fish. The intensity of utilization of its stocks is throughout high. The stocks, the volume of catch of pike-perch and its contribution to total catch of fish in the southern reservoirs are higher than in the northern, which can be seen from the data presented below:

Water body	Average yearly catch, ton	% of average annual catch of fish
Ivankov Reservoir	2.9	1.2
Uglich Reservoir	1.9	0.8
Gorkii Reservoir	17.9	4.2
Cheboksar Reservoir	2.8	1.5
Kuibyshev Reservoir	352.6	7.8
Volgograd Reservoir	444.1	13.5
Tsimlyansk Reservoir	1170.0	9.1
Lake Il'men'	98.0	3.9

Intensive fishing exerts appreciable influence on the population structure of pike-perch in some of the freshwater bodies starting from such index as the range of the age series. Thus, in Lake Il'men' cases were recorded with catches of pike-perch in the age up to 19 years. Already in the 1960s, in this lake individuals with age up to 12 and in 1970s up to 10+–11+ were reported [Fedorova, 1974; Sakharov, 1980]. However, in the 1980s the limiting range of the age series shortened up to 7+ [Kuderskii, Vetkasov and Kartsev, 1985]. In reservoirs the utilizations of the stocks of pike-perch is less intense. In view of this, the age series in the populations of pike-perch appears to be more prolonged. In Ivankov and Uglich reservoirs, despite lower stocks, individuals with age up to 12 years have been reported [Kuderskii and Nikanorov, 1983]; the figures for Chaboksar are up to 13 [Zaloznykh, 1985b], Volgograd up to 15 [Nebol'sina, Elizarova and Abramova, 1980] and for Tsimlyansk up to 18+.

The intensive exploitation of the stocks variously affects the individual pike-perch populations. This is clearly seen from the change of duration of the age series of pike-perch in three reservoirs according to five-year periods (Table 41). In Kuibyshev Reservoir, for the four incomplete five-year periods, the average range of the age series in pike-perch population increased from 10.3 to 11.8 years, while in Volgograd Reservoir it slightly decreased from 11.0 to 10.2 years. At the same time, in pike-perch population of Tsimlyansk Reservoir, the number of age classes sharply decreased: from 15.8 in 1968–1972 to 10.6 in 1983–1987.

According to the average multiyear data, the rate of decrease in the number of individuals of pike-perch is higher than that of the ichthyomass (Table 42).

In view of this, the number of individuals in pike-perch populations by the age of mass maturation decreases to 10/13–5/177. Ichthyomass in reservoirs, in which its culmination does not coincide with mass maturation, decreases with reference to the culminating age class to 10/11–5/11. In individual reservoirs the rates of decrease of both indices fluctuates in the following manner. Number of individuals decreases in Gorkii, Cheboksar and Volgograd reservoirs and in Lake Il'men' to 5/172, 10/207, 10/147 and 5/11 respectively; ichthyomass decreases to 5/7, 10/11, 5/11 and 5/6 respectively.

Table 41. Average duration of age series according to five-year periods in pike-perch populations from reservoirs

Reservoir	1968-1972	1973-1977	1978–1982	1983–1987
Kuibyshev	10.3	9.6	9.8	11.8
Volgograd	11.0	8.6	10.6	10.2
Tsimlyansk	15.8	16.2	12.8	10.6

Note: Data for Kuibyshev Reservoir are from 1970–1972, that for Volgograd Reservoir from 1969–1972.

Table 42. Rate of decrease of the number of individuals and ichthyomass of age groups in pike-perch populations from reservoirs

Age	Cheboksar Reservoir, average for 1981–1987		Volgograd Reservoir, average for 1969–1987		Tsimlyansk Reservoir, average for 1964–1987	
	Population	Ichthyomass	Population	Ichthyomass	Population	Ichthyomass
1+–2+	3.5	0.8	2.0	0.9	0.4	0.2
2+–3+	2.1	0.6	1.6	0.9	0.8	0.4
3+–4+	1.6	0.8	1.5	0.9	1.3	0.9
4+–5+	1.8	1.1	3.2	2.2	2.0	1.4
5+–6+	1.6	1.04	2.1	1.6	1.9	1.4
6+–7+	1.7	1.3	2.0	1.5	1.7	1.3
7+–8+	1.8	1.5	1.8	1.2	1.9	1.6
8+–9+	2.3	2.0	4.0	3.2	1.8	1.4
9+–10+	2.1	2.2	1.0	1.7	1.8	1.5
10+–11+	—	—	—	—	1.7	1.6
11+–12+	—	—	—	—	1.4	1.2
12+–13+	—	—	—	—	1.1	1.2
13+–14+	—	—	—	—	1.8	1.4
14+–15+	—	—	—	—	1.7	1.6
15+–16+	—	—	—	—	1.8	1.7

It was observed above that in bream populations by the onset of mass maturation ichthyomass of age groups decreased on the average to 5/16. In the four above listed populations this is not observed since in them ichthyomass culmination and mass maturation coincide in time. The observed differences may be attributed to the peculiarities of the structure of bream and pike-perch populations. The effect of fishing in the given case is of secondary importance, which is clearly confirmed by the data for Lake Il'men'. In bream population of this lake, by the onset of mass maturation, the ichthyomass decreases to 10/53 while in pike-perch to 5/6. Both species of fish are included in fishery of Lake Il'men' at high intensity.

The age of ichthyomass culmination in pike-perch populations from the eight reservoirs under review, according to the multiyear average data, varies from 3–3+ to 6–6+ (Table 43). In each population the age of ichthyomass culmination often changes from year to year (Fig. 11). In pike-perch population from Lake Il'men', ichthyomass culmination in individual years was observed in age classes from 3–3+ to 5–5+, in Kuibyshev Reservoir, from 4–4+ to 8–8+ and in Volgograd Reservoir from 1–1+ to 4–4+. the average age of ichthyomass culmination is also variable in individual populations (Table 44).

Table 43. Number of individuals and ichthyomass of age groups in pike-perch populations from reservoirs and Lake Il'men'

Age	Ivankov Reservoir, 1976–1982		Uglich Reservoir, 1971–1982		Gorkii Reservoir, 1981–1987		Cheboksar Reservoir, 1981–1987	
	Thousand	Ton	Thousand	Ton	Thousand	Ton	Thousand	Ton
1–1+	33.5	0.8	6.7	0.2	212.1	6.6	429.6	12.6
2–2+	37.5	3.2	12.4	1.6	86.0	6.9	121.7	16.1
3–3+	20.5	4.1	12.5	3.5	25.7	7.9	58.4	26.6
4–4+	11.8	5.8	13.0	4.6	15.3	12.0	36.5	33.7
5–5+	8.4	7.1	10.3	7.5	6.0	8.9	20.8	30.6
6–6+	8.0	9.3	8.1	9.6	2.2	4.8	14.3	30.4
7–7+	3.5	5.6	2.0	3.3	1.0	6.2	8.6	23.9
8–8+	1.0	2.1	1.4	2.8	0.4	1.5	4.8	16.5
9–9+	1.6	4.6	0.9	2.3	0.2	0.8	2.1	8.4
10–10+	0.3	1.1	0.5	1.7	0.3	0.3	1.0	3.9
11–11+	0.1	0.2	0.1	0.3	0.03	0.1	0.1	0.5
12–12+	0.2	0.5	0.1	0.2	—	—	—	—
13–13+	—	—	—	—	—	—	—	—
14–14+	—	—	—	—	—	—	—	—
15–15+	—	—	—	—	—	—	—	—
16–16+	—	—	—	—	—	—	—	—
17–17+	—	—	—	—	—	—	—	—
18–18+	—	—	—	—	—	—	—	—

Contd

Table 43 Contd.

Age	Kuibyshev Reservoir, 1970–1987		Volgograd Reservoir, 1969–1987		Tsimlyansk Reservoir, 1964–1987		Lake Il'men, 1979–1984	
	Thousand	Ton	Thousand	Ton	Thousand	Ton	Thousand	Ton
1–1+	—	—	4.25	603.3	275.1	48.5	1.08	65.0
2–2+	161.2	31.7	2.18	688.4	739.5	274.3	1.04	175.6
3–3+	371.7	148.6	1.37	740.0	923.4	617.8	1.10	397.7
4–4+	372.6	337.7	0.92	831.5	712.3	719.2	0.50	328.3
5–5+	265.6	354.1	0.29	385.5	354.5	520.3	0.22	272.8
6–6+	176.9	249.7	0.14	237.8	190.0	363.7	0.05	91.5
7–7+	76.7	146,8	0.07	156.4	110.0	281.8	0.01	9.1
8–8+	41.7	83.8	0.04	132.7	58.0	177.8	—	—
9–9+	22.4	65.9	0.01	42.1	32.7	123.5	—	—
10–10+	9.2	30.9	0.01	25.3	17.8	81.7	—	—
11–11+	5.5	22.4	0.004	29.3	10.6	56.3	—	—
12–12+	1.3	6.7	0.002	8.2	7.4	46.5	—	—
13–13+	0.5	2.3	—	—	6.8	39.0	—	—
14–14+	—	—	—	—	3.7	27.9	—	—
15–15+	—	—	—	—	2.2	17.9	—	—
16–16+	—	—	—	—	1.2	10.6	—	—
17–17+	—	—	—	—	1.1	10.0	—	—
18–18+	—	—	—	—	2.0	1.0	—	—

Note: See note to Table 38.

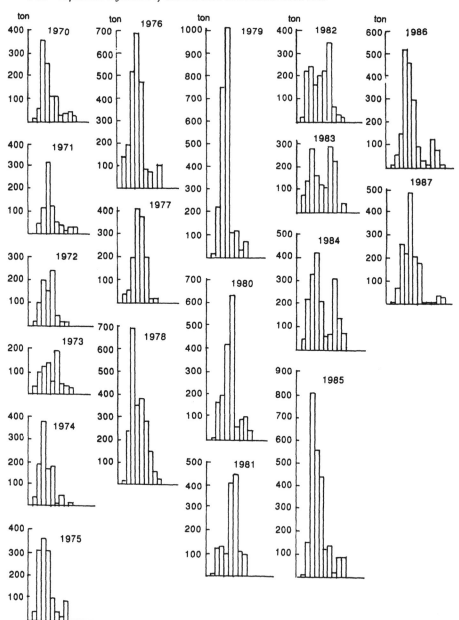

Fig. 11. Dynamics of ichthyomass of age groups in populations of pike-perch from Kuibyshev Reservoir. Age class 2+ shown in the extreme left.

The age of ichthyomass culmination is the highest in pike-perch population of Kuibyshev Reservoir and lowest in Volgograd Reservoir. The average age of ichthyomass culmination evidently does not depend on fishing at the existing level of its intensity. Thus in Kuibyshev, Volgograd and Tsimlyansk reservoirs and Lake Il'men', the extent of utilization of stocks of pike-perch is high, which is reflected even in such an index as the range of age series. In the years 1983–1987, in reservoirs, the average range of the age series was 10.2–11.8 years while in Lake Il'men' it was not more than eight years. However, the average age of ichthyomass culmination in Lake Il'men' and two reservoirs was 3.1–3.7 years while in the third reservoir, i.e. Kuibyshev, it was 5.4 years. The age of ichthyomass culmination is, for the most part, linked with the harvest level of individual pike-perch generations.

The age of ichthyomass culmination in pike-perch populations often changes not randomly but with a definite periodicity. In this respect, pike-perch differs from Volga zander, which is closer to it not only in systematic relationship but in many aspects of biology. The periodicity of change of the age of ichthyomass culmination is very well expressed in the populous populations of pike-perch from Kuibyshev, Volgograd and Tsimlyansk reservoirs and Lake Il'men'. In Volgograd Reservoir, over a period of 19 years, one four-year period (1974–1977) and three two-year periods were observed. In Lake Il'men' despite the restricted age series (7+) of pike-perch, over a period of six years, one three-year and one two-year periods were observed. The most vivid periodicity of

Table 44. Age of ichthyomass culmination in pike-perch populations from reservoirs and Lake Il'men' (number of years)

Water body	Age of ichthyomass culmination								Average age of ichthyomass culmination	Number of years of observation
	1–1+	2–2+	3–3+	4–4+	5–5+	6–6+	7–7+	8–8+		
Lake Il'men'	—	—	3	2	1	—	—	—	3.7	6
Gorkii Reservoir	—	—	1	6	—	—	—	—	3.9	7
Cheboksar Reservoir	—	—	3	2	1	1	—	—	4.0	7
Kuibyshev Reservoir	—	—	—	5	6	3	2	2	5.4	18
Volgograd Reservoir	3	2	4	10	—	—	—	—	3.1	19
Tsimlyansk Reservoir	—	3	8	9	4	—	—	—	3.6	24

change of the age of ichthyomass culmination is seen from the example of pike-perch population of Kuibyshev and Tsimlyansk reservoirs:

Year	Age of ichthyomass culmination		Year	Age of ichthyomass culmination	
	Kuibyshev Reservoir	Tsimlyansk Reservoir		Kuibyshev Reservoir	Tsimlyansk Reservoir
1964	—	5+	1976	5+	3+
1965	—	2+	1977	5+	3+
1966	—	3+	1978	4+	4+
1967	—	4+	1979	5+	3+
1968	—	5+	1980	6+	4+
1969	—	5+	1981	7+	4+
1970	4+	4+	1982	8+	3-1
1971	5+	4+	1983	8+	3+
1972	6+	4+	1984	5+	2+
1973	7+	4+	1985	4+	3+
1974	4+	4+	1986	5+	2+
1975	4+	5+	1987	6+	3+

Particularly, in pike-perch from Kuibyshev Reservoir, out of 18 years of observations such a periodicity was noticed in 14 years. This population had one each of the three-year (three underlines), four-year (single underline), and five-year (double underline) period in addition to one two-year period (broken underline). The lower number of longer periods (only one four-year period shown with a single underline) is characteristic of the change of age of ichthyomass culmination in pike-perch population from Tsimlyansk Reservoir. But here one comes across five two-year periods (shown with broken underline) and on the whole for 24 years of observation the periodicity was recorded for 14 years.

The periodicity in the change of the age of ichthyomass culmination is one of the forms of reflection of the leading position of individual generations for many years in the structure of populations, which differ from the neighboring generations in the level of harvest and rate of decrease of ichthyomass with age. I have mentioned this point, though briefly, a number of times before. It is also confirmed by the following comparison of the ichthyomass of age class 4+, in which, in pike-perch population from Kuibyshev Reservoir, the leading role of individual generations with culminating ichthyomass* is observed:

Year	Ichthyomass of age class 4+, ton	Ichthyomass culmination (ton) and age		Year	Ichthyomass of age class 4+, ton	Ichthyomass culmination (ton) and age	
1970	360.0	360.0	(4+)	1979	750.4	1,034.4	(5+)
1971	114.4	311.2	(5+)	1980	144.8	641.6	(6+)
1972	201.6	234.4	(6)	1981	132.8	448.8	(7+)
1973	120.0	189.0	(7+)	1982	235.2	346.8	(8+)
1974	379.2	379.2	4+	1983	279.0	288.8 - 8+	
1975	356.8	356.8	(4+)	1984	326.4	424.8 - 5+	
1976	515.0	668.0	(5+)	1985	809.0	809.0	(4+)
1977	200.0	408.0 - 5+		1986	149.7	521.0	(5+)
1978	693.6	693.6	(4+)	1987	261.0	487.6	(6+)

It will be seen from the data presented above that the 1966 pike-perch generation occupied leading position in ichthyomass for four years, from 1970 to 1973, since the three generations following it had a lower harvest. Their ichthyomass on attaining the age of 4+, as compared to the 1966 generation was 5/16, 5/9 and 1/3 respectively. Precisely likewise the 1981 generation in 1985, at the age of 4+, was leading in ichthyomass, and retained this position in 1986 and 1987 since the harvest of the two generations following it was lower. At the age of 4+, the ichthyomass of the 1981 generation was higher than that of the next two generations by 5.4 and 3.1 times.

Besides the above relatively simple cases, in the pike-perch population from Kuibyshev Reservoir, a more complex version was also noticed. The 1971 generation, at the age of 4+, in 1975 had an ichthyomass which was 5/7 of the 1972 generation following it. Nevertheless the first generation occupied the leading position for two years —1975 and 1976. Similarly the 1974 generation in 1978, in the age of 4+, had an ichthyomass 10/11 of the one neighboring it but 4.8, 5.2 and 3.0 times as high as the three generations following it. As a result of such a combination of ichthyomass of four generations following the 1974 generation, the latter retained its leading position for five years 1978–1982. Thus to retain the leading position in respect of ichthyomass for several years in a row and for the development of periodicity in the change of age of ichthyomass

---

* For greater precision, the periods of change of the age of ichthyomass culmination discussed above are circumscribed by boxes.

culmination, only the high harvest of individual generation is not sufficient. It is necessary to have a definite combination of the harvest levels of many neighboring generations which ensure one of them, for several years in a row, a leading position in ichthyomass. At the same time, it is not difficult to conclude that a prolonged leading position of one generation is the consequence of instability of the reproduction of stocks.

In pike-perch populations from the reservoirs in question and Lake Il'men' complexity of their structure due to ichthyomass subculminating age groups is expressed weakly. It appears only in individual years and reflects the dissimilar level of harvest of individual generations.

Mass maturation of pike-perch populations is observed in the age of 4–4+ to 6–6+. If one considers only the reservoir populations, one conclusion may be drawn about variation in this index from north to the south. In Ivankov and Uglich reservoirs, mass maturation in pike-perch is observed in the age class 6–6+, in the remaining Volga reservoirs in the age of 5–5+. At the same time, in pike-perch population from the southernmost of the investigated reservoir—Tsimlyansk—mass maturation sets in the age classes 4–4+. A similar age of mass maturation is also found in pike-perch population from Lake Il'men'. Hence it may be said that zonal variation of the age of onset of mass maturation is not a linear function in populations of pike-perch. It is apparently associated not only with geographic but also historic factors. However, a discussion of this problem falls beyond the purview of the present book.

From the multiyear average data in different pike-perch populations the number of mature (starting from the age. of mass maturation) age classes is 4–15. Their lowest number (4) is found in pike-perch from lake Il'men', and the highest in Tsimlyansk Reservoir (15). In pike-perch from Kuibyshev and Volgograd reservoirs, their maximum number is respectively 9 and 8, in the remaining four reservoirs 7.

Considerable variation in the number of mature age classes in pike-perch from different water bodies may be considered as an indicator of neutrality of this index in relation to the total number of individuals in a population. This conclusion must be borne in mind while developing exploitation regimes for stocks of pike-perch in the general perspective of the production potential of water bodies.

The correlation of the age of ichthyomass culmination and mass maturation is an important index determining the conformity of the structure of population with one of the three types. According to the multiyear average data, the pike-perch populations under reference fall into two groups in respect of this index. In the first group, ichthyomass culmination occurs one year before mass maturation. In the second group, both the events occur concurrently. The first group includes populations of pike-perch from Gorkii, Cheboksar, and Volgograd reservoirs and Lake Il'men', while the second group includes populations from Ivankov, Uglich, Kuibyshev and Tsimlyansk reservoirs.

In view of the observed heterogeneity in different population of pike-perch, the ratio of ichthyomass of immature and mature (starting from the age of mass maturation) age classes* is not constant. In majority of water bodies, the ichthyomass of immature individuals is less and constitutes 2/5th to 9/10th of mature individuals. Only in pike-perch from Gorkii and Volgograd reservoirs is the ichthyomass of immature age classes higher than that of mature ones, as much as 1.2 and 2.2 times respectively.

In contrast to the multiyear average data, material on multiyear variation in the correlation of the age of ichthyomass culmination and mass maturation reveals a more complex picture (Table 45). In pike-perch populations from four reservoirs, the age of ichthyomass culmination, in individual years, is found to be lower or higher than the age of mass maturation, or coincides with it. On the other hand, in pike-perch from Gorkii and Volgograd reservoirs, in all years of observation, ichthyomass culmination preceded the onset of mass maturation.

It will be seen from Table 45 that in all water bodies (except Kuibyshev Reservoir), in pike-perch populations ichthyomass culmination often precedes mass maturation than follow it. Hence it must be recognized that the structure of population of this species often corresponds to the third than to the second type. This conclusion, at first glance, seems unexpected, as in relation to pike. Based on the ecological specificities of pike-perch one would presume that the structure of its populations would correspond to the second type. However, factual material does not confirm this intuitive representation.

CATFISH

The structure of catfish populations is examined on the material obtained from Tsimlyansk Reservoir. Here catfish is a secondary fishing object. In individual years, its catches reached 1102 ton or 10.6% the total catch [Kuderskii and Dronov, 1984]. During 1981–1985, the average annual catch of catfish was 251 ton or 2% of the total catch of fish from the reservoir.

Catfish belongs to the category of fish with long life cycle. The population of this species from Tsimlyansk Reservoir includes up to 30

---

* The ichthyomass of immature age classes was determined without taking into account fish in the age of 0+ and 1+ in view of their incomplete estimate. Hence the value is somewhat low. However, such a calculation does not change the essence of ovarall conclusions.

Table 45. Correlation of ages of ichthyomass culmination and mass maturation in pike-perch populations (number of years)

Water body	Age of ichthyomass culmination in relation to age of mass maturation			Ratio of ichthyomass of immature to mature fish	Number of years of observation
	Less	Equal	More		
Lake Il'men'	3	2	1	0.8	6
Gorkii Reservoir	7	—	—	1.2	7
Cheboksar Reservoir	5	1	1	0.7	7
Kuibyshev Reservoir	5	6	7	0.5	18
Volgograd Reservoir	19	—	—	2.2	19
Tsimlyansk Reservoir	11	9	4	0.4	24

age classes. In the range of age series catfish surpasses all the above examined fish. Despite perennial fishing the age series in the catfish population persistently included a large number of age classes; in the recent years, their number has risen in comparison with the early 1960s:

Five year period	1963–1967	1968–1972	1973–1977	1978–1982	1983–1987
Average number of age classes	19.0	23.0	28.0	28.0	28.0

Such an age composition of the catfish population can hardly be considered rational. The data presented here puts in doubt the conclusion about the need for intensification of catches of this fish in order to reduce losses inflicted on stocks of other commercial species of fish.

The age of ichthyomass culmination in the catfish population fluctuates in a wide range, from 4 to 23 completed years, which was not observed for even any other species of fish. The strict periodicity of change of the age of ichthyomass culmination also seems rather unusual:

For the 27 years of observations only three periods of change of the index under reference were noticed: 16-year period (1961–1976), four year period (1977–1980) and seven-year period (1981–1987). The last period is still operative. A distinct periodicity of change in the age of ichthyomass culmination is due, firstly, to the rare appearance of high harvest generations and, secondly, to such combination of the harvest level of neighboring generations, as would ensure one of the generations to retain its leading position for a long period of time. In other words, in catfish population, periodic changes of the age of ichthyomass culmination are due to the same causes as for population of pike-perch; but

Year	Age of ichthyomass culmination	Year	Age of ichthyomass culmination	Year	Age of ichthyomass culmination
1961	8	1970	17	1979	9
1962	9	1971	18	1980	10
1963	10	1972	19	1981	4
1964	11	1973	20	1982	5
1965	12	1974	21	1983	6
1966	13	1975	22	1984	7
1967	14	1976	23	1985	8
1968	15	1977	7	1986	9
1969	16	1978	8	1987	10

their effect is manifested more evenly, primarily because of the low harvest of majority of generations.

The above noted periodicity may be considered as a proof basically of the unfavorable conditions for reproduction of stocks of catfish in Tsimlyansk Reservoir. Only in rare years do conditions establish for the appearance of high-harvest generations. As the results of observations show, the condition favoring appearance of high-harvest generations of catfish in Tsimlyansk Reservoir after 1956 appeared only three times. It is not difficult to see that, at frequent appearance of generations commensurate in harvest, the duration of leading position for each generation often would be absolutely different; and with frequent appearance of such generations, the observed periodicity could be even disrupted, as was the case in populations of other species of fish.

In view of the long life cycle and presence in the catfish population of many (up to 30) age classes, its structure in individual years becomes complex because of the appearance of ichthyomass subculminating age groups (Table 46). The latter could be younger as well as older. In individual years, in catfish population, there are up to two younger and three older subculminating age groups.

In catfish population of Tsimlyansk Reservoir, mass maturation is observed in the age of six completed years. In view of such early mass maturation with overall range of age series a predominant part of ichthyomass of catfish is concentrated in the mature age classes, which is clearly seen from the data presented in Table 46. It must be noted that such a predominance of ichthyomass of mature fish is not characteristic of even one of the above examined species of fish. Herein lies one of the peculiarities of the structure of catfish population.

Thanks to the prolonged age series, for the 27 years of observations 11–25 mature (starting from mass maturation) age classes were noticed in catfish population. In the last five-year periods (starting from 1973),

Table 46. Number of individuals and ichthyomass of age groups in catfish population of Tsimlyansk Reservoir for individual years

Age	1964 In thousand	1964 Ton	1970 In thousand	1970 Ton	1976 In thousand	1976 Ton	1980 In thousand	1980 Ton
3	104.8	171.9	54.2	88.9	16.7	27.4	21.8	357.6
4	103.9	305.6	35.7	105.0	16.9	49.8	14.0	41.2
5	69.0	303.8	67.0	294.8	55.4	243.6	9.4	41.3
6	57.3	344.3	97.7	587.4	114.7	689.5	25.8	155.3
7	58.6	453.5	49.4	459.7	25.5	197.4	7.9	60.8
8	50.2	480.9	35.8	342.9	19.3	185.2	5.8	55.3
9	73.4	843.6	17.3	199.2	9.5	108.9	19.0	218.9
10	65.8	855.8	10.2	136.9	6.4	86.8	43.1	580.0
11	129.4	2,003.8	12.8	198.7	17.2	266.0	10.1	156.0
12	77.1	1,356.1	13.5	236.8	28.0	492.9	7.9	140.0
13	30.9	610.2	13.1	258.1	15.7	309.8	4.0	80.0
14	7.5	164.7	13.0	286.0	8.6	190.3	3.0	66.6
15	6.4	156.6	12.1	296.1	4.6	111.6	8.2	199.6
16	3.6	97.6	21.4	574.9	2.9	78.8	13.1	351.9
17	34.6	1,018.1	56.1	1,654.2	5.2	154.1	7.0	205.3
18	0.9	28.5	29.0	931.3	5.4	172.5	3.7	118.3
19	0.1	4.2	4.5	156.8	5.0	174.5	1.8	63.2
20	—	—	—	—	4.7	174.7	1.1	41.8
21	—	—	—	—	3.7	149.8	1.9	77.9
22	—	—	—	—	6.6	282.8	1.7	73.6
23	—	—	7.4	335.6	16.6	748.7	1.5	69.8
24	—	—	—	—	6.9	329.3	1.2	56.1
25	—	—	—	—	0.3	13.0	0.9	42.9
26	—	—	—	—	—	—	1.3	68.3
27	—	'	—	—	—	—	2.7	146.3
28	—	—	—	—	—	—	0.8	42.7
29	—	—	—	—	0.3	19.7	—	—

Note : See note to Table 29.

on the average, its population shows 22 such classes. Earlier I have repeatedly noted the neutral nature of such an index which is the number of mature age classes responsible for the level of harvest of generations. This conclusion is clearly confirmed by the data on catfish population. Despite large number of mature age classes, a feature unknown in other species of fish examined in this book, a majority of catfish generations are distinguished by low harvest. Hence, in relation to commercial fish as a whole, the problem of developing principles of optimization of age series of their population is quite valid.

The age of ichthyomass culmination in catfish population, as a rule, is higher than the age of mass maturation. In 24 out of 27 years, it was higher than age of mass maturation by 1–17 years, and in these years the structure of population corresponded to the second type. Considering the range of the age series and early mass maturation in relation to the limiting age, such a correlation of both indices must be considered fully natural. However, in individual years, a deviation is noticed from the strict correspondence to the second type characteristic for catfish population structure. Twice the age of ichthyomass culmination preceded the period of mass maturation and once coincided with it. The observed exception is associated with the appearance of the high-harvest 1977 generation, which surpassed the neighboring generations in ichthyomass and occupied the leading position in the population from 1981–1987. The ichthyomass of three-year olds of the 1977 generation was higher than that of the six generations following it by as much as 2.8, 3.1, 2.6, 3.1, 2.6, and 2.9 times respectively. Such a rare combination of harvest levels of several neighboring generations caused a change from the second type for a period of three consecutive years.

Despite deviation from the norms characteristic of catfish population, the ichthyomass of the mature part of the population of these above mentioned three years was much higher than that of the immature one. The ratio of ichthyomass of immature age classes to that of mature ones was 0.3, 0.5 and 0.2 respectively. In other words, in the ratio of ichthyomass of these two parts of the population, even in these exceptional years, the structure of population is of the second type. Hence the unusual correlation for the ages of ichthyomass culmination and mass maturation of the three consecutive years does not alter the overall conclusion and correspondence of the structure of catfish population to the second type.

DISCUSSION

In this chapter I have examined material on a limited number of water bodies—eight large reservoirs and one large lake. Moreover, for the majority of species of fish information is not available for all the investigated water bodies; particularly for blue bream and catfish, the information was extracted from only one reservoir. In view of this, the possibilities for a conclusive comparative analysis, based on the material presented in this chapter, are limited than that in the previous chapter. Considering the above fact as also the aim of the present book, which is mostly concerned with analysis of the quantitative indices, in the overview part concluding this chapter, it would be expedient to discuss certain aspects of such an index as the ichthyomass.

The population dynamics of commercial fish as a scientific problem is quite multifaceted in nature and includes wide range of aspects and concepts. Among these ichthyomass is one such important index. All fish production estimates ultimately lead to this index. The results of fishing are expressed in ichthyomass units. The studies conducted on the quantitative estimates of fish in water bodies contain conclusive information also in the form of ichthyomass.

There are individual nodal points in fish populations. These, to a great extent, determine their structure. Some of these points are associated with the concept of ichthyomass. Among them, mention may be made of the age of ichthyomass culmination, its correlation with the age of mass maturation, and so on. Below, an attempt is made to examine individual aspects connected with the concept of ichthyomass in relation to the population.

One such aspect is the change of ichthyomass between the age of its culmination and age of mass maturation of fish. Data relating to this aspect (multiyear average data) are presented in Table 47. These need the following explanation: in the Table, I have shown the ratio of ichthyomass of two age classes, viz. 1) that having the maximum ichthyomass, and 2) that reaching maturity. In population where ichthyomass culmination precedes mass maturation cf fish (populations of third type), the reported ratio signifies the factor of decrease of ichthyomass of age classes from the first event to the second. In cases where ichthyomass culmination succeeds mass maturation of fish (populations of the second type), the ratio indicates the factor of increase of ichthyomass of age classes from the moment of mass maturation to ichthyomass culmination. In case both events coincide in time, the ratio of ichthyomasses equals unity.

An analysis of the data presented in Table 47 first of all highlights the predominance of populations belonging to the third type. All populations of bream fall in this category. It also includes the lone population of blue bream, four out of five populations of roach, three out of five populations of pike, one out of three populations of Volga zander and four out of eight populations of pike-perch. It thus follows that

Table 47. Change of ichthyomass from its culmination to the age of mass maturation (from multiyear average data)

Water body	Bream	Roach	White bream	Blue bream	Pike	Volga zander	Pike-perch	Catfish
Ivankov Reservoir	5.3	1.1	(2.6)	—	1.5	—	1.0	—
Uglich Reservoir	1.9	(1.7)	(4.9)	—	1.9	—	1.0	—
Gorkii Reservoir	2.1	3.3	—	—	1.0	—	1.4	—
Cheboksar Reservoir	2.3	2.2	—	—	1.3	—	1.1	—
Kuibyshev Reservoir	1.8	—	—	—	(2.6)	1.0	1.0	—
Saratov Reservoir	3.9	—	—	—	—	—	—	—
Volgograd Reservoir	2.3	1.6	(1.6)	—	—	1.7	2.2	—
Tsimlyansk Reservoir	1.4	—	(2.8)	1.4	—	1.0	1.0	(2.2)
Lake Il'men'	5.4	—	—	—	—	—	1.2	—

Note: Figures in parentheses denote the increase of ichthyomass from the age of mass maturation to its culmination in population of the second type. For catfish, average data are reported for 1961–65. The years of collection of material for the remaining species are given elsewhere in the text.

ichthyomass culmination occurs before mass maturation; it is a widespread phenomenon among commercial fish of the water bodies under examination.

What are the possible reasons for this phenomenon? Since fishing is developed in all the reservoirs mentioned in Table 47, in some cases even intensively, it becomes imperative to, first of all, evaluate its possible influence on the ratio of ichthyomass of the identified two age classes of fish. It must be at once mentioned that fishing does not play a decisive role in determining the nature and scale of the phenomenon reflected in Table 47. Fish of culminating age classes in the populations of the third type, according to the operative laws of fishing, have as yet not reached the commercial level. In view of this, their possible catches are limited in two respects: commercial scale and percentage catch of young (precisely speaking, undersize fish). In connection with these limitations, an attempt is made to determine the mesh size of stationary, and dragnets and the gear itself. Moreover, in all the water bodies mentioned herein, the catches of valuable fish (bream, pike-perch and some others) are permitted within limits of the annually established limits. While specifying these limits due attention was paid to the basic peculiarities of the structure of fish populations including their size and age.

The intensity of exploitation of commercial stocks of fish listed in Table 47 is not uniform. Roach catches throughout are much lower than their possible limit. Stocks of bream, too, are exploited nonuniformly. In view of this, it would be expedient to conduct the following comparison. In the intensively exploited bream population from Lake Il'men', the ichthyomass from the age of its culmination to mass maturation decreases to 5/27. An almost similar decrease (10/53) is observed in the weakly exploited bream population from Ivankov Reservoir. In Saratov Reservoir, bream stocks are exploited not to the fullest extent. However, the ichthyomass for the above noted period decreased to 10/39, while in the intensively exploited bream population from Tsimlyansk Reservoir it decreased only to 5/7. In the recently commissioned Cheboksar Reservoir, in which fishing is in the initial stages of establishing, the ichthyomass of bream population from the age of culmination to mass maturation decreased to 10/23. Such a decrease of ichthyomass is noticed also in Volgograd Reservoir, where bream stocks are exploited adequately fully.

The stocks of pike-perch and pike are utilized, as a rule, intensively. But in populations of these fish one does not observe such a large decrease of ichthyomass as in roach and bream. If this decrease were to be mainly related to the intensity of fishing, one would expect, in pike-perch and pike, the maximum decrease of ichthyomass in the age classes under discussion. However this does not happen, particularly because throughout a strict regime of fishing is observed, which is established at all fishing establishments.

The principal factors causing a decrease of ichthyomass, that too considerable, of age classes from its culmination to mass maturation, have an ecological character. Leading among them are the ones which determine the number of individuals (and ichthyomass) of younger age classes of fish up to reaching mass maturity. These include the efficiency of reproduction, availability of food in early stages of development and, what is most important for understanding the problem under discussion, the presence of adequate food resources for various young age groups. Relationships of the "predator-prey" type have a serious implication. The determining role of ecological factors becomes manifest in the repeatedly stated cases of periodic changes of the age of ichthyomass culmination. Very often, the rate of weight gain plays the leading role. At very high individual weights the proportion of ichthyomass of one and the same age class is found to be different. The age of attaining mass maturation also affects the ratio of ichthyomass of culminating age groups and of ones that have reached mass maturity. This factor needs to be considered during a comparison of populations of one and the same species from different water bodies.

From the ecological point of view, one can also explain the variation in the structure of populations of many species of fish. Though from the multiyear average data the structure of a population belongs to the third type, in individual years it may transform into the second type. Such a phenomenon is noticed in individual populations of practically all the species of fish considered above. The structure of blue bream population is marked by very high variation. In 10 out of 20 years of observation its population corresponded to the third type and in 6 years, to the second type (cf. Table 37).

Variation in the population structure in relation to the correlation of the age of ichthyomass culmination and mass maturation complicates the overall picture of the dynamics of fish populations and confirms the inadequacy of a single time collection of material. For an exhaustive solution of these problems one requires not the annual "sections" of the population structure but multiyear observations on its variation.

Besides the populations under discussion, Table 47 also includes populations in which (from multiyear average data) the ichthyomass of age classes after attaining mass maturity does not decrease but increases. In them, the age of ichthyomass culmination is higher than the age of mass maturation. This group includes all populations of pike-perch, the lone population of catfish, one (of the five) population of roach and one (of the four) population of pike. The observed phenomenon occurs despite uninterrupted fishing exerting its influence on that part of the population which has crossed the threshold of maturity. The above noted confirmation about the leading role of ecological factors in determining the peculiarities of the age dynamics of ichthyomass is applicable to this group of populations to a much greater extent than to the group considered earlier.

Among the factors facilitating the formation of populations of the second type, mention must also be made of the lower rate of elimination in the younger age classes. Accordingly, the rate of decrease of the number of individuals with age decreases and a large number of fish reach the middle and old age, thereby increasing the total ichthyomass of the respective age classes. No less important is the degree of food availability since at limited food availability, firstly, there is a decrease in the growth of mass of fish and secondly, the intensity of elimination increases. Quite often, the rate of growth of ichthyomass plays a decisive role. A higher growth of mass compensates for, to a great extent, the loss of ichthyomass due to natural mortality, increase the total ichthyomass of individual age classes, and changes its ratio.

Populations of pike-perch (four), Volga zander (two) and pike (one) occupy a special position in Table 47. In these populations, according to the multiyear average data, the ratio of ichthyomass of culminating age classes and reaching mass maturation is one, since both events coincide in time. As the analysis of data on these populations for individual years shows that in some years their structure conforms to the second type and in others to the third type. This is generally decided by the relationship of the harvest levels of the neighboring populations. In other words, in the ultimate analysis this version of the change of ichthyomass is associated with definite ecological factors.

The age of mass maturation divides fish populations into two unequal parts. One of them (young) includes immature age classes and the other (starting from the age of mass maturation) mature age classes*. The question arises, what is the ratio of the ichthyomass of these two qualitatively different groups of individuals? Material relating to this topic was obtained from the multiyear average data presented in Table 48.

All populations of fish species in question, in the ratio of the ichthyomass of immature and mature individuals, fall into two groups. The first group includes populations in which the ichthyomass of immature age classes is higher than that of mature ones. The ratio of two

---

* In such a division, among immature fish, there will be some which have matured (in age classes neighboring age of mass maturation) and, on the other hand, among mature fish there would be immature individuals. I subscribe to this simplification, since a strict division complicates the estimates but does not add principal changes in the results of discussion.

Table 48. Ratio of ichthyomass of immature and mature age classes (from multiyear average data)

Water body	Bream	Roach	White bream	Blue bream	Pike	Volga zander	Pike-perch	Catfish
Ivankov Reservoir	9.8	0.6	0.05	—	1.3	—	0.9	—
Uglich Reservoir	3.7	0.1	0.01	—	1.5	—	0.9	—
Gorkii Reservoir	5.2	2.6	—	—	0.2	—	1.5	—
Cheboksar Reservoir	4.3	2.1	—	—	0.5	—	0.8	—
Kuibyshev Reservoir	2.9	0.6	—	—	0.02	0.4	0.5	—
Saratov Reservoir	3.3	—	—	—	—	—	—	—
Volgograd Reservoir	3.0	—	0.13	—	—	1.2	2.8	—
Tsimlyansk Reservoir	0.9	—	0.03	0.9	·	0.3	0.4	0.2
Lake Il'men'	13.9	—	—	—	—	—	0.9	—

Note: Data on catfish are for the year 1987.

ichthyomasses in this group of population is more than one. The second group includes populations with predominance of ichthyomass of mature individuals. The ichthyomass ratio in this case is less than one.

Earlier, while discussing the population structure of the individual fish species, I repeatedly mentioned the ratio of ichthyomass of immature and mature age classes. However, the results presented in Table 48 are somewhat unusual. First of all, our attention is caught by the predominance of ichthyomass of immature individuals almost in all populations of bream, except one. Even in such reservoirs as Ivankov, Cheboksar, and Saratov, in which the degree of exploitation of bream stocks is not high, the ichthyomass of immature fish is 3.3–9.8 times as high as that of mature fish. Predominance of the ichthyomass of mature age classes is observed only in bream population from Tsimlyansk Reservoir despite the fact that here bream fishing is most intensive. Thus in most bream populations, including the ones that are weakly exploited, a unique predominance of immature age classes is observed, not only in the number of individual but also in their ichthyomass.

In the predominance of ichthyomass of immature individuals bream has no parallel among other species of fish from the reservoirs in question. It may be said that, in conditions of large reservoirs in the plains and such unique lakes as Lake Il'men', bream shows unusually high reproduction potential of its stocks, which has not been fully evaluated in the investigations conducted on this topic.

In contrast to bream, in all populations of white bream, the dominant part of the ichthyomass is concentrated in the mature age classes. This may also be considered unexpected since white bream in many aspects behaves as bream. It also behaves like bream in small lakes.

In catfish population a large part of ichthyomass is persistently and stably shared by mature fish. The same cannot be said about blue bream. According to the multiyear average data, its population shows predominance of ichthyomass of mature fish. However, in view of the instability of reproduction of stocks and appearance of high harvest generations, in individual years, in blue bream immature fish could predominate in their ichthyomass.

In roach, pike, white bream and pike-perch, in some populations, there is a predominance of immature individuals, in others it is mature. Moreover variation in the ratio of ichthyomass of these two groups is possible in individual years. However, in the course of this work it was not possible to analyze this variation in detail. It may be noted that several populations of roach differ considerably from those in small lakes in respect of the ichthyomass of immature individuals.

Usually attention is not paid to the variation in the ratio of two ichthyomasses in one and the same species as also in its single population but in different years. But this fact needs to be considered not only in the theoretical analysis but in determining the possible catches as well as practical exploitation of stocks of commercial fish.

The ichthyomass of immature and mature age classes of fish is an important population-structure index. It permits a clear understanding of correlations in different populations of the processes of growth and development. In literature similar data are seldom examined. However, this is not due to negation to their importance in fish production and fishing studies but to the extreme paucity of quantitative data characterizing different aspects of the population structure. Gradual accumulation of material on the absolute (complete) estimates of the population and (primarily) of ichthyomass in water bodies will, in the near future, permit removal of this lacuna. At the same time, availability of such data ensures a more objective evaluation of the fish production phenomena and working out the optimum exploitation regimes of live resources of water bodies. Together with quantitative material characterizing the structure of fish populations, data on the ichthyomass of different age classes, including the total figures for immature and mature fish are also equally important. Their significant importance is clearly revealed from the discussion about the data presented in Table 48.

Both mature and immature age classes are the essential constituent elements of population. In the former, there is an accumulation of ichthyomass and the process of development of the reproductive system. Mature fish ensure reproduction of population. It is because of reproduction of population as distinct from generation, that long-term super-organismal formations become established. In natural conditions the ratio of ichthyomass of immature and mature age classes establishes under the impact of concrete ecological conditions. It is not strictly determined and changes in wide limits (Table 48). Under conditions of regulated fishing, where figures are released annually and which are revised depending on the needs, there is no necessity to strictly hold on to any one scheme of ratio of ichthyomasses studied and approximated on the example of an isolated water body or their small groups. In this respect, it would be expedient to hold on to a specific approach to water bodies and the object of fishery.

More realistic is the approach, which would permit regulation of the ichthyomass of several immature age classes in each water body according to the availability of food to them and to more older age groups of fish, and to regulate ichthyomass of mature individuals in limits as would ensure guaranteed reproduction of populations to the extent, allowing the most complete utilization of the bioproduction potential of the water body. Methods of ichthyomass regulation of different age classes should be based taking into account the ecological peculiarities of different species of fish, their adaptation potential, as well as specificities of those aquatic ecosystems of which these fishes are a part.

Such is the production aspect of possible regulation of the population structure of commercial fish. However, this is not the only possible application. Other aims and objectives may be for instance, consumer-related, ecological, recreational, or may relate to the maintenance of quality of the water medium. In accordance with each of these aims, the directed measures on regulating the structure of fish populations would be different. What remains unchanged is the need to collect and analyze material on the absolute estimates of fish in water bodies and, based upon that data, to develop quantitative characteristics of different population-structure indices.

As has been repeatedly stated earlier, ichthyomass culmination is an important index. It is examined in two aspects. One of them is the age of ichthyomass culmination. The other is its value and attendant role in the formation of ichthyomass of population as a whole.

The age of ichthyomass culmination has been examined in detail in sections dealing with individual fish species. Generalizing the material presented above, it becomes essential to consider the following general problems.

The age of ichthyomass culmination in the populations of one and the same fish species changes in wider limits (Table 49). According to the multiyear average data, in six species of fish, for which the material is available from more than one water body, the maximal age exceeds the minimal 1.5–5 times. The maximum range of change of this index is observed in bream populations where it is as much as 5.0 times. In the remaining fish species it varies from 1.5–3.00 times. Still greater variation of the age of ichthyomass culmination is observed during analysis of data for individual populations for different years. For instance, in bream populations this age, in individual years, varies from 1–1+ to 11–11+. In populations of other species of fish, the year to year variation of this feature is less. In white bream the maximum age does not exceed 9–9+, in pike perch 8–8+, roach and blue bream 7–7+, pike and Volga zander 6–6+. However, the maximum year by year variation of the age of ichthyomass culmination has been recorded in catfish. In the only population studies in this respect from Tsimlyansk Reservoir this age varied from 4 to 23 completed years [Kuderskii and Dronov, 1984].

The wide local (in populations from different water bodies) and year to year variation of the age of ichthyomass culmination in one and the same species of fish is due to many ecological factors. These factors can be divided into three groups determining respectively, the efficiency of reproduction, the pressure of elimination of young age groups and the rate of weight gain. In different populations, the role of each of these

Table 49. Age of ichthyomass culmination and mass maturation in populations of commercial fish from reservoirs and Lake Il'men' (from multiyear average data)

Water body	Beam	Roach	White bream	Blue bream	Pike	Volga zander	Pike-perch	Catfish
Ivankov Reservoir	$\frac{4\text{-}4+}{9\text{-}9+}$	$\frac{3\text{-}3+}{4\text{-}4+}$	$\frac{5\text{-}5+}{3\text{-}3+}$	—	$\frac{3\text{-}3+}{4\text{-}4+}$	—	$\frac{6\text{-}6+}{6\text{-}6+}$	—
Uglich Reservoir	$\frac{6\text{-}6+}{9\text{-}9+}$	$\frac{6\text{-}6+}{4\text{-}4+}$	$\frac{4\text{-}4+}{3\text{-}3+}$	—	$\frac{3\text{-}3+}{4\text{-}4+}$	—	$\frac{6\text{-}6+}{6\text{-}6+}$	—
Gorkii Reservoir	$\frac{4\text{-}4+}{8\text{-}8+}$	$\frac{2\text{-}2+}{4\text{-}4+}$	—	—	$\frac{3\text{-}3+}{3\text{-}3+}$	—	$\frac{4\text{-}4+}{5\text{-}5+}$	—
Cheboksar Reservoir	$\frac{3\text{-}3+}{8\text{-}8+}$	$\frac{2\text{-}2+}{4\text{-}4+}$	—	—	$\frac{2\text{-}2+}{3\text{-}3+}$	—	$\frac{4\text{-}4+}{5\text{-}5+}$	—
Kuibyshev Reservoir	$\frac{6\text{-}6+}{9\text{-}9+}$	—	—	—	$\frac{5\text{-}5+}{3\text{-}3+}$	$\frac{5\text{-}5+}{5\text{-}5+}$	$\frac{5\text{-}5+}{5\text{-}5+}$	—
Saratov Reservoir	$\frac{6\text{-}6+}{9\text{-}9+}$	—	—	—	—	—	—	—
Volgograd Reservoir	$\frac{4\text{-}4+}{7\text{-}7+}$	$\frac{2\text{-}2+}{3\text{-}3+}$	$\frac{6\text{-}6+}{3\text{-}3+}$	—	—	$\frac{1\text{-}1+}{3\text{-}3+}$	$\frac{4\text{-}4+}{5\text{-}5+}$	—
Tsimlyansk Reservoir	$\frac{5\text{-}5+}{6\text{-}6+}$	—	$\frac{4\text{-}4+}{3\text{-}3+}$	$\frac{4\text{-}4+}{5\text{-}5+}$	—	$\frac{4\text{-}4+}{4\text{-}4+}$	$\frac{4\text{-}4+}{4\text{-}4+}$	$\frac{-}{6\text{-}6+}$
Lake Il'men'	$\frac{2\text{-}2+}{10\text{-}10+}$	—	—	—	—	—	$\frac{3\text{-}3+}{4\text{-}4+}$	—

Note: The age of ichthyomass culmination is shown in the numerator; the age of mass maturation is shown in the denominator. For catfish, the age of ichthyomass culmination is not given due to large fluctuations of the value from year to year.

factors differs. For instance, in catfish population the harvest level of individual generations and, consequently, the factors determining it are of decisive importance in determining year by year variation. The high-harvest generations retain the leading position in ichthyomass for many years, and their aging causes a consequently change of the age of ichthyomass culmination. The effect of relationship of harvest levels of neighboring generations on year by year variation of the index in question is also observed in other fish, to which attention was drawn when the structure of their populations was described. However, it is not manifest in such distinct form as in catfish.

While examining all cases of change in the age of ichthyomass culmination in connection with the harvest level of individual generations the following need to be verified. In essence, the reference must be made not to the values of harvest but to the relative stability of reproduction of stocks over a period of many consecutive years. With stable efficiency of this process the range of variation in the age of ichthyomass culmination reduces at consistently low as well as consistently high harvests. With fluctuating efficiency of reproduction, the range of variation of the age in question increases. With sharply unstable reproduction, this index acquires the extreme form as in the catfish population.

The year by year variation in the age of ichthyomass culmination sometimes appears as periodicity of change of this index. The latter phenomenon was examined in detail when describing the structure of populations of individual species of fish. Hence there is no need to discuss it again.

Besides the above mentioned variation in the age of ichthyomass culmination, attention must be paid to some parallelism with the degree of manifestation of the phenomenon in question. For instance, in populations of bream and pike-perch from Lake Il'men', the age of ichthyomass culmination is lower than that in reservoirs. In Cheboksar Reservoir, in populations of roach and pike-perch, this age is the same as that in Gorkii but in populations of bream and pike it is lower. In population of the bream, pike and pike-perch of Kuibyshev Reservoir, the age of ichthyomass culmination is higher than that in Gorkii and Cheboksar reservoirs. Some other parallelism of such type can also be mentioned. However the material presented in Table 49 is not sufficient for a more detailed discussion of the problem of independent interest.

The age of ichthyomass culmination divides the populations of fish into two parts. In age classes preceding culmination, there is increase of ichthyomass. Upon attaining the maximum value (culmination), there occurs a decrease (gradual or with independent repeated rises) of ichthyomass of age class through to zero. Hence ichthyomass culmination occupies the threshold position between the two parts of a population, distinguished by fish production. However, in the population-structure analysis, of sole importance is not the boundary position of the culminating ichthyomass in the age series. Of great interest is the percentage fraction of culminating age class in the total population ichthyomass. The data relating to this problem are practically wanting in literature. Material available at my disposal has been presented in Table 50.

A comparison of data presented in Tables 49 and 50 allows us to conclude that the contribution of the culminating ichthyomass does not depend on the type of structure of population (from multiyear average data). In bream, whose all populations belong to the third type, the culminating ichthyomass accounts for 16.0–26.6% of the total population ichthyomass. In white bream with second type of structure of its population, the culminating ichthyomass constitutes 14.0–29.9% of the total.

The proportion of culminating ichthyomass does not depend on the range of the age series. Also in catfish, whose life span reaches 30 years, and in blue bream, whose life span is less than half of catfish, the percentages of the culminating ichthyomass are rather quite close. The same may be said also during a comparison of other parts of fish species (pike-perch and blue bream, bream and blue bream, pike and roach, and so on). If eventually some small noncorrespondences arise, on the whole, they may not have a decisive significance.

The most important thing, which attracts attention during analysis of the data presented in Table 50, is the high contribution of culminating ichthyomass in most populations. It is possible to consider such values as somewhat unexpected, taking into account the range of the age series in many species of fish. Culminating ichthyomass is a unique wave in populations which finds its expression in the harvest level of individual generation. The high contribution to total populational ichthyomass distinguishes it among the neighboring age classes and plays a significant role in the fish production processes.

It is not difficult to visualize from the data presented above that such an index as ichthyomass culmination unveils additional prospects for a more objective solution to problems of exploitation of the stocks of individual species of fish as well as production potential of aquatic ecosystems as a whole. It can be used as a starting point of estimates for working out the optimal regimes, from the production viewpoint, for fishing in inland water bodies.

Table 50. Ichthyomass culmination in % of populational ichthyomass (from multiyear average data)

Water body	Bream	Roach	White bream	Blue bream	Pike	Volga zander	Pike-perch	Catfish
Ivankov Reservoir	21.8	22.2	18.1	—	32.6	—	21.0	—
Uglich Reservoir	22.9	24.5	20.3	—	36.8	—	25.0	—
Gorkii Reservoir	16.0	37.5	—	—	23.4	—	21.4	—
Cheboksar Reservoir	16.0	32.7	—	—	23.8	—	16.5	—
Kuibyshev Reservoir	16.0	—	—	—	27.3	34.0	23.9	—
Saratov Reservoir	23.1	—	—	—	—	—	—	—
Volgograd Reservoir	20.7	19.6	14.0	—	—	32.6	21.4	—
Tsimlyansk Reservoir	17.8	—	29.4	22.7	—	25.8	21.0	23.2
Lake Il'men'	26.6	—	—	—	—	—	29.7	—

Note: Data for catfish relates to the year 1970.

CONCLUSION

The fish population of a water body, or ichthyocenose, is represented by the totality of directly or indirectly interacting populations, each of which is made up of individuals of one species. Populations are dynamic formations, which is confirmed by the material presented in the two chapters of this book. High dynamism, that is, variation, engulfs the diverse population structure indices and sometimes reaches considerable ranges. The structure of populations of one and the same species of fish, inhabiting lakes and reservoirs, often appear heterotypic and, as a rule, subject to year by year changes. However, against the background of high variation, there appear some features which are common to large groups of populations. These could serve as key characteristic for building a scheme of classification.

In ichthyology, the separation of populations into types based on the range of age series, that is, life span (short, medium and long life cycles), is widely used. G.N. Monastyrskii (1952) identified types of spawning populations of fish from the correlation of values of replenishment and surplus. U.E. Riker (1979) formulated the concept of the types of ideal populations of fish based on a comparison of natural and fishing mortality. There are many such examples. It must be borne in mind that any classification must conform to definite fundamental requirements. First, not the external (formal) but internal features expressing the orderly links of the phenomena in question should form the basis of separating a set of objects into types. Second, the classification should be adequate for the problem for whose analysis it is being used.

I have divided the populations of commercial fish into three types based on the correlation in the age of ichthyomass culmination and mass maturation [Kuderskii, 1983, 1984, 1986]. While developing this classification I considered that fish populations are formations that are complex in structural aspect. They include many diverse individuals of one species in which, first of all, distinct same-age classes are identifiable, which correspond to different generations.

The similar age group of fish begins its existence from the spawn and with growth changes from one age class into the other, right up to disappearance from the water body as a result of mortality upon reaching the limiting age. There occurs a parallel change in such important population indices as the number of individuals in a generation and its ichthyomass. The nature of their change with age also differs. The number of individuals in an age group (generation) decreases with the growth of fish from the maximum characteristic of the first stages of development to zero by the time of disappearance of the generation (mortality due to natural causes or fishing). Orderly changes in the number of individuals in a generation with age of fish is graphically represented by the concave mortality curve [Zasosov, 1976]. Unlike the number of individuals, the ichthyomass of a generation at first increases, attains at a definite age the maximum value (ichthyomass culmination) and then either there is a gradual or rapid decrease down to zero. The curve of change of biomass of a generation with age is bell-shaped with two minima—in the right (initial) and left (end) parts. The position of maximum on this curve varies for different species of fish and even for different populations of one and the same species.

In works on the dynamics of fish stocks it is customary to analyze the course of changes in the population of a generation with age. Considerably lesser attention is paid to the analysis of peculiarities of age dependent changes of ichthyomass of generations. Moreover such population index as ichthyomass of age groups carries in it specific information about the production processes which have great relevance in the analysis of the problem of commercial exploitation of populations. In particular, the age at which ichthyomass culmination occurs as if "divides" the generation into two unequal parts. In years preceding the age of ichthyomass culmination, there occurs an increase of biomass and it continuously increases despite the intensively on-going reduction in population on account of natural mortality. In this period of life cycle, the equally effective process "reduction of population—increase of mass of individuals" has a positive value. Hence years of life of a fish from the moment of spawning to attaining maximum ichthyomass by the generation may be considered as the productive part of the life cycle.

Upon attaining culmination, ichthyomass of the generation starts decreasing. Moreover, the route of its decrease depends on the species of fish as well as the specific habitat of the population. In this period of life of the fish, unlike the earlier, there is a simultaneous decrease of population and biomass of age groups. The continuing individual weight-gain by fish appears insufficient to offset the loss of ichthyomass due to natural (and commercial) mortality. Hence the production process for this part of the generation, on the whole, occurs with a negative sign.

The change of ichthyomass with an increase in the growth of fish is one of the important characteristics typical of populations. The value of maximum ichthyomass reached in the process can serve as an index of luxuriance of population in specific habitat conditions.

Besides the magnitude of ichthyomass, each generation of fish has its specific age of maturation. It is usually different in different species. Moreover, in individual populations of the same species, the age of maturation may show some changes*. Maturation ensures the accomplishment of the vital phenomenon—reproduction of population.

Like ichthyomass culmination the age of maturation divides the life cycle of a generation into two unequal parts. During the first period fish are immature. They only gradually prepare themselves for the ensuing reproduction of numbers (and also of ichthyomass) of population. This period is marked by intensive weight-gain by fish and development of their gonads. The rate of their development, to a greater extent, depends on the conditions of the habitat in which the population lives [Koshelev, 1984]. The age of maturation is a unique type of turning point in the life of a generation. From this moment the second period of the life cycle begins, in which the generation carries out the functions of reproduction of number of individuals in a population.

Thus in fish of the same generation two important processes take place with age, which are important for the fate of the population. First, the increase of ichthyomass and second, preparation for and entry into the phase of ensuing reproduction of the number of individuals (and ichthyomass). Both these phenomena are of definite importance for blooming populations and their serious disruption could lead to the degradation of populations.

The increase of ichthyomass and maturation in the same generation are accomplished, to a known degree, independently; although at individual stages of life cycle they may be seen to follow a parallel course. During the initial years of life of fish, there is an increase in the ichthyomass of generation and maturation of gonads. However, upon reaching the age of ichthyomass culmination and mass maturation, the direction of both phenomena seems to be heterotypic: biomass of the generation decreases but the surviving fish continue to fulfill the function of producers of the population. Relative independence of both processes may be seen from the well known fact: maturation of fish occurs unavoidably despite different magnitudes of the ichthyomass of generations varying, as shown above, in wide limits.

Both events—attaining the maximum values of ichthyomass and onset of mass maturation—in the life of a generation could coincide in time. However, instances are available, when the points of ichthyomass culmination and mass maturation are separated from each other by one

---

* Maturation of fish of one generation often occurs over a period of many years. However in the present case for a generalized examination of the structure of populations this aspect is not of that fundamental significance.

or several years. Then it would be possible to separate the third part of life cycle of a generation, which includes age classes between the culmination of ichthyomass and onset of mass maturation. A simultaneous consideration of each of the three parts of life cycle of fish leads to the conclusion that the fate of a generation is determined, first of all, by the peculiarities of the course of the processes of decrease of population, growth of ichthyomass and development of the reproductive system in the period up to the onset of such critical moments, as ichthyomass culmination and mass maturation.

The discussion above relates to individual generations of fish. However, investigators as well as fishery personnel, as a rule, do not deal with individual generation but with their totalities or populations. In the latter, all generations of the same species may occur simultaneously, the appearance and conservation of which is possible in specific habitat conditions. Because of this, the overall picture of the dynamics of populations acquires new qualitative dimensions. The latter can be seen at least from the fact that a generation is an ephemeral formation. It appears during multiplication of fish and then either gradually or rapidly (depending on the type of fish) disappears from the reservoir. Moreover, the duration of existence of a generation depends on the life span of the longest surviving individuals. Unlike generation, a population exists for a long time, experiencing only more or less significant fluctuations in the number of individuals and ichthyomass and accordingly changes of structure under the influence of fluctuations of the conditions of existence.

By analogy with generations in populations, too, one identifies the age at which the ichthyomass reaches the maximum value (culmination) and the age of mass maturation of fish. These two events mostly determine the peculiarities of the structure of specific populations.

Analysis of the correlation of the age of ichthyomass culmination and mass maturation makes it possible to divide populations of commercial fish according to the peculiarities of their structure into three types (Fig. 12).

The simplest in character are populations of fish belonging to the first type (Fig. 12a). In them, ichthyomass of age groups continuously increases with age and attains culmination at mass maturation. After participation in spawning, mature fish die either entirely or in a large number. The multiplying producers, in the populations belonging to the first type, are either absent or are few in number. Hence, the ichthyomass of large groups, after their mass maturation and more so upon reaching culmination, decreases sharply or falls to zero.

In populations of the first type, the production phenomena are most simple, since these are not complicated by the presence of additional

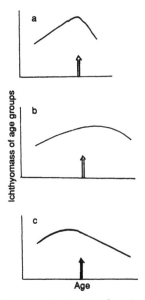

Fig. 12. Correlation of the ages of ichthyomass culmination and mass maturation (shown by arrow) (Explanation given in the text).

age groups following the ones that reached mass maturity and also affecting the reproduction of the fish stock. Dynamics of the population structure and its ichthyomass, both as a whole and according to age groups, also become simple. From the moment of appearance of larvae to attaining maturity in such populations, two parallel processes occur which are of primary importance for the production phenomena. These are the increase of ichthyomass and the development of the reproductive system.

The first type is characteristic of the structure of populations of fish such as smelt, pink salmon, chum and other Far Eastern salmons, smaller forms of cisco and others, and for lamprey among cyclostomes. In the present book these species have not been considered.

Populations whose structure conforms to the second type differ significantly from those with the first type of structure. The difference is manifest in the divergence of the age of ichthyomass culmination and the onset of mass maturation. As can be seen from Fig. 12b, in populations of the second type, ichthyomass culmination is observed after fish reach mass maturity. Hence ichthyomass culmination can occur in the age group in which individuals have repeatedly participated in reproduction.

Populations of commercial fish belonging to the second type differ by the great complexity of their structure. Three following heterogeneous groups can be identified among them:

— age classes from the time of appearance of larvae to mass maturation;
— age classes from attaining mass maturity to onset of ichthyomass culmination;
— age classes older than the age of ichthyomass culmination.

The role of each of these groups is different in the functioning of the populations. In age classes belonging to the first group, ichthyomass increases and maturation of fish occurs. In the second group, ichthyomass continues to add but fish comprising the group simultaneously take part in reproduction of the population. Finally, fish of the third group continue to perform function of reproduction of population but the ichthyomass of these groups constantly decreases despite constant weight-gain by isolated individuals.

The production phenomena in populations of commercial fish of the second type are complicated since, firstly, the increase of populational biomass continues even after attaining mass maturity by fish right up to the onset of ichthyomass culmination. Secondly, in the reproduction of population two types of age classes participate. The first type includes age classes that increase the populational biomass and the second type includes age classes which are older than the culminating age groups and which reduce the populational biomass.

In populations belonging to the second type, the major part of the populational ichthyomass is contributed by age groups older than the age of onset of mass maturation. This second type substantially differs from the populations of first type in which predominant part of ichthyomass is contributed by immature age groups of fish. The observed difference in the "attraction" of ichthyomass of population to immature groups in one case and to mature groups in the other finds a special reflection in the peculiarities of fish production phenomena in each of these types of populations.

Populations of some species of fish with structure conforming to the second type have been examined in both chapters.

Among populations of commercial fish there are some in which ichthyomass culmination occurs before the onset of mass maturation (Fig. 12c). These populations belong to the third type. This type of population is observed in many species of fish inhabiting small lakes as well as reservoirs in plains with extensive zone suitable for fish reproduction.

Among the above identified types of populations of commercial fish the third type draws our attention, since in its structure it appears rather unusual from the point of view of conventional concepts established in ichthyology. In this type the population ichthyomass increases only in those age groups which are represented by immature fish. Ichthyomass culmination also occurs in immature age groups of fish. By the time of onset of mass maturation the value of ichthyomass of age groups reduces considerably.

Characteristic peculiarity of populations of the third type lies in that in them, unlike in the second type, the predominant part of ichthyomass is concentrated in the immature and partly mature age groups of fish. At the same time, the ichthyomass of that part of the population, which includes fish that have attained maturity or fully mature individuals, is relatively small.

The structure of populations of the third type is no less complex than the second type. In particular, in populations of the third type, three parts differing from each other can be identified. The first part includes fish from the larval stages of development to ichthyomass culmination. These are almost entirely immature. Fish of these age classes ensure increase of populational ichthyomass and simultaneous development of the reproductive system.

The second part includes age classes from culmination of ichthyomass to reaching mass maturity. Fish from this part of the population are partially mature but the ichthyomass of these age classes shows a decrease. In view of this, the role of the second part in the functioning of the population is quite unique: on the one hand, there is a decrease (at times quite considerable) of the populational ichthyomass and, on the other, fish comprising this part only begin (and not fully) to participate in reproduction of population.

The third part of population comprises age classes older than those that attained mass maturity. Their ichthyomass in comparison to the culminating ichthyomass is greatly reduced and continues to decrease with age. Age classes comprising this third part ensure reproduction of population, and they do not increase but decrease the populational · ichthyomass.

The described peculiarity of structure of populations belonging to the third type is often attributed to the effect of the intensity of fishing. Supporters of this viewpoint do not allow for the possibility of existence of populations of fish for which predominance of immature individuals would be typical. However, in the numerous examples of populations of the third type examined by me, the number of mature fish was low (in terms of ichthyomass) not because they were not caught but because of the peculiar dynamics of their stocks.

During analysis of populations of many species of fish in chapters 1 and 2, attention was repeatedly paid to the following phenomenon. In some species of fish, in conditions of some reservoirs or in some years, populations of second type are formed; in other years the populations are of the third type. Such cases are not isolated. In view of this, it may be considered that the structure of fish populations can be fairly labile. This happens because the structure arises as the end result of interactions of potential possibilities provided by fish themselves as well as consequence of specific conditions of the habitat. This phenomenon has great importance in practice of fixing the fishing regimes: these should be expediently established not for individual species of fish as a whole, but applied to individual populations. For individual commercially important fish, the fishing regime should be worked out together with regular forecasts of catches in order to consider the on-going changes not only in the magnitude of stocks but also in the structure of populations. Accordingly, it will be possible to resolve the problem of most complete utilization of live resources and conservation of stocks from depletion.

Analysis of the dynamics of ichthyomass of age groups, besides separating the above mentioned three types of populations, makes it possible to objectively discuss the fish production phenomena on the basis of populations. It must be mentioned that ichthyomass of population is made of three components, viz. ichthyomass of immature, partly mature, and fully mature age classes. The role of each of these components in the fish production process is not equal. In age classes comprising immature fish, in all the three types of populations, there is an increase of ichthyomass and the latter, on the average, is higher in each subsequent age class in comparison to the preceding age class. Despite the on-going reduction of population in this period of life cycle, the intensity of weight-gain is such that it not only compensates for the loss of ichthyomass due to natural mortality but even ensures its considerable increase in each subsequent age class.

The role of partly mature age classes in the populations of the first and second types is similar. In these cases, the weight-gain by fish is also higher than losses occurring due to natural causes as a result of elimination of some individuals. However, in the populations of the third type the situation is different. The partially mature classes only sometimes add to the ichthyomass. At other times, there may be a fall in their ichthyomass. The latter is clearly seen from Fig. 12c, which schematically shows that starting from the age of ichthyomass culmination its decrease occurs, although in succeeding age classes (in the age of mass maturation) not all fish reach maturity. Within limits of these age classes, the weight-gain by fish does not compensate for loss of ichthyomass due to natural mortality.

The role of age classes comprising mature fish in the formation of ichthyomass is not similar in different types of populations. In populations of the first and third types, in age classes comprising mature fish, a loss of populational ichthyomass is noticed despite continuing weight-gain by individuals which have escaped elimination. A particularly sharp change of populational biomass of mature fish occurs, as mentioned above, in populations of the first type. In populations of the second type, age classes comprising mature fish partly increase the ichthyomass (up to its culmination) and partly lose it (after attaining culmination). Moreover, with multiage structure of the mature part of the population, the increase of ichthyomass is obtained in a small number of age classes.

Thus, in all the three types of population, in age classes comprising mature fish, there is a reduction of the populational ichthyomass with only occasional increase (partly in populations of the second type). However, these age classes of fish perform another role in population, viz., they ensure reproduction of population (and of ichthyomass). Accordingly, they play a decisive role in ensuring the very existence of populations.

In conclusion, it must be mentioned that the above identified types of populations of fish make it possible to fully evaluate the fish production phenomena and, on its basis, work out and implement the necessary organizational measures to ensure optimum use of live resources, development, and observance of scientifically based regimes of fishing in inland reservoirs.

BIBLIOGRAPHY

Anonymous. 1982. Izmenenie struktury rybnogo naseleniya evtofiruemogo ozera (Change in the Structure of Fish Population of a Lake Undergoing Eutrophication). Nauka, Moscow, 248 p.

Anonymous. 1986. Issledovanie vzaimosvyazi kormovoi bazy i ryboproduktivnosti na primere ozer Zabaikal'ya (Investigation of the Interaction of the Food Base and Fish Productivity as Demonstrated by the Example of Trans-Baikal Lakes). Nauka, Leningrad, 232 p.

Balagurova, M.V. 1970. O vozrastnykh, sezonnykh i godovykh razlichiyakh v pitanii okunya (On the age-related, seasonal and year-by-year differences in the feeding of perch). In *Vodnye Resursy Karelii i Puti Ikh Ispol'zovaniya*, Petrozavodsk, Karelia, pp. 354–374.

Bandura, V.I. and S.V. Shibaev. 1984. Chislennost', produktsionnye vozmozhnosti i rol' plotvy v e'kosisteme Gor'kovskogo vodokhranilishcha (Population, production potential and the role of roach in the ecosystem of the Gorkii Reservoir). In *Sb. Nauchn. Tr. GOSNII Ozer. i Rech. Ryb. Khoz-va*, No. 222, pp. 103–109.

Baranova-Filon, V.V. 1980. Vliyanie rybolovstva na zapasy plotvy Ivankovskogo i Ugilichskogo vodokhranilishch (The effect of fishing on the stocks of roach in the Ivankov and Uglich reservoirs). Ibid, No. 145, pp. 52–59.

Berg, L.S. 1939. Obzor rybnogo naseleniya melkikh ozer Leningradskoi oblasti (Review of fish population of small lakes of Leningrad Region). *Izv. Vsesoyuzn. NII Ozer. i Rech. Ryb. Khoz-va*, Vol. 22, pp. 7–13.

Burmakin, E.V. 1960. Rezul'taty rabot po obezryblivaniyu dvukh ozer Leningradskoi oblasti (Results of works on fish removal from two lakes of Leningrad Region). *Nauchn. -Tekhn. Byul. Gos. NII Ozer. i Rech. Ryb. Khoz-va*, No. 12, pp. 20–24.

Burmakin, E.V. 1961. Absolyutnaya chislennost' okunya i ego ikhtiomassy v malykh ozerakh (Absolute numbers of perch and its ichthyomass in small lakes). In *Tr. Soveshch. po Dinamike Chislennosti Ryb*, Izd-vo AN SSSR, Moscow, pp. 235–237.

Burmakin, E.V. and L.A. Zhakov. 1961. Opyt opredeleniya ryboproduktivnosti okunevogo ozera (Experiment to determine fish productivity of a perch lake). *Nauch.-Tekhn. Byul. Gos NII Ozer. i Rech. Ryb. Khoz-va*, Nos. 13–14, pp. 27–28.

Chikova, V.M. 1966. Sostoyanie nerestnykh stad i razmnozhenie ryb v Cheremshanskom i Suskanskom zalivakh Kuibyshevskogo vodokhranilishcha (The state of spawning stocks and reproduction of fish in Cheremshan and Suskan bays of Kuibyshev Reservoir). *Tr. Inta Biologii Vnutrennykh Vod AN SSSR*, No. 10(13), pp. 29–45.

Chugunov, N.L. 1928. Biologiya molodi promyslovykh ryb Volgo-Kaspiiskogo raiona (Biology of juveniles of commercial fish of Volga-Caspian Region). *Tr. Astrakh. Nauchn. Rybokhoz Stantsii*, Vol. 6, No. 4, pp. 3–282.

Domrachev, P.F. 1922. K voprosu klassifikatsii ozer Severozapadnogo kraya (On the problem of classification of lakes of the Northwestern Territory). *Izv. Ros. Gidrol. In-ta*, No. 4, pp. 1–43.

Dronov, V.G. 1974. Biologiya i promysel soma Tsimlyanskogo vodokhranilishcha (On the biology and fishery of catfish of Tsimlyansk Reservoir). *Tr. Volgogr. Otd. Gos NII Ozer. i Rech. Ryb. Khoz-va*, Vol. 8, pp. 161–175.

Dronov, V.G. 1982. Razmernaya i vesovaya struktura populyatsii soma Tsimlyanskogo vodokhranilishcha (The size and age structure of catfish populations of Tsimlyansk Reservoir). *Sb. Nauchn. Tr. Gos NII Ozer. i Rech. Ryb. Khoz-va*, No. 184, pp. 49–53.

Dryagin, P.A. 1947. Ob opredelenii potentsial'nogo rosta i potentsial'nykh razmerov ryb (On the determination of potential growth and potential sizes of fish). *Izv. VNII Ozer. i Rech. Ryb. Khoz-va*, Vol. 26, No. 1, pp. 111–121.

Fesenko, G.M. 1976. O ratsional'nom ispol'zovanii zapasov sintsa Tsimlyanskogo vodokhranilishcha (On the rational use of stocks of zope in Tsimlyansk Reservoir). *Tr. Volgogr. Otd-niya Goss NII Ozer. i Rech. Ryb. Khoz-va*, Vol. 10, No. 2, pp. 23–29.

Fedorova, G.V. 1974. Biologiya i dinamika chislennosti sudaka ozera Il'men' (Biology and population dynamics of pike-perch from Lake Il'men'). *Izv. Gos NII Ozer. i Rech. Ryb. Khoz-va*, Vol. 86, pp. 73–89.

Gerd, S.V. 1949. Nekotorye zoogeograficheskie problemy, izucheniya ryb Karelii (Some zoogeographic problems of studying fish in Karelia). In *Prirodnye Resursy, Istoriya i Kul'tura Karelo-Finskoi SSR*, Gosizdat Karelo-Fin. SSR, Petrozavodsk, No. 2, pp. 110–115.

Gladkii, G.V. 1964. Osobennosti khishchincheskogo pitaniya okuni i shchuki v razlichnykh tipakh ozer Narochanskoi gruppy (Characteristics of predation of perch and pike in different types of Naroch' group of lakes). In *Biologicheskie Osnovy Rybnogo Khozyaistva na Vnutrennykh Vodoemakh Pribaltiki*, pp. 75–78.

Gorbunova, Z.A., A.M. Gulyaeva and Yu.S. Ditrenko. 1978. Ikhtiofauna Veshkelitskikh ozer (Ichthyofauna of Veshkelitsy lakes). *Tr. Sev. Nauch.- Issled. i Proektno-Konstrukt. In-ta Ryb. Khoz-va*, Vol. 8, pt. 1, pp. 62–72.

Gulin, V.V. 1968. Nekotorye voprosy metodiki opredeleniya otnositel'nogo vozrastnogo sostava promyslovogo stada ryb na vnutrennykh vodoemakh (Some problems relating to the method of determination of the age composition of commercial stocks of fish from inland reservoirs). *Vopr. Ikhtiologii*, Vol. 8, No. 1(48), pp. 139–153.

Gulyaeva, A.M. 1951. Materialy po biologii okunya Onezhskogo ozera (Data on the biology of perch of Lake Onega). *Tr. Karelo-Fin. Otd-niya VNII Ozer. i Rech. Ryb. Khoz-va*, Vol. 3, pp. 150–168.

Kitaev, S.P. 1984. E'kologicheskie osnovy bioproduktivnosti ozer raznykh prirodnykh zon (Ecological bases of bioproductivity of Lakes of different natural zones). Nauka, Moscow, 232 p.

Korablev, I.P. 1972. Organizatsiya rybolovstva (Organizing fishing). *Tr. Tatar. Ot-niya Gos. NII Ozer. i Rech. Ryb. Khoz-va*, No. 12, pp. 180–200.

Koshelev, B.V. 1984. E'kologiya razmnozheniya ryb (Reproductive Ecology of Fish). Nauka, Moscow, 309 p.

Kuderskii, L.A. 1962. O rybokhozyaistvennom znachenii okunya v ozerakh Karelii (On the fishery significance of perch in lakes of Karelia). *Nauch. -Tekhn. Byul. Gos NII Ozer. i Rech. Ryb. Khoz-vo*, No. 15, pp. 47–50.

Kuderskii, L.A. 1983. Kul'minatsiya ikhtiomassy vozrastnykh grupp u promyslovykh ryb vnutrennykh vodoemov i strategiya rybolovstva (Ichthyomass culmination of age groups of commercial fish of inland reservoirs and the fishing strategy). *Ryb. Khoz-vo*, No. 7, pp. 41–43.

Kuderskii, L.A. 1984. Tip struktury populyatsii promyslovykh ryb i strategiya ispol'zovaniya ikh zapasov (The type of population structure of commercial fish and the strategy of exploitation of their stocks). *Sb. Nauch. Tr. Gos NII Ozer. i Rech. Ryb. Khoz-va*, No. 211, pp. 109–117.

Kuderskii, L.A. 1986. Tip populyatsii promyslovykh ryb (The type of population of commercial fish). In *Dinamika Chislennosti Promyslovykh Ryb*, Nauka, Moscow, pp. 231–245.

Kuderskii, L.A., Yu.V. Aleksandrov and L.G. Perminov. 1968. Vozrasty kul'minatsii ikhtiomassy i massovoi polovozrelosti v populyatsii leshcha malykh ozer Pskovskoi oblasti (Age of ichthyomass culmination and mass maturation in bream population of small lakes of Pskov Region). *Sb. Nauch. Tr. Gos NII Ozer. i Rech. Ryb. Khoz-va*, No. 283, pp. 142–156.

Kuderskii, L.A., S.A. Vetkasov and V.N. Kartsev. 1985. Kul'minatsiya ikhtiomassy v vozrastnykh i razmernykh gruppakh leshcha i sudaka ozera Il'men' (Ichthyomass culmination in age- and size-groups of bream and pike-perch of Lake Il'men'). Ibid, No. 237, pp. 31–49.

Kuderskii, L.A. and V.G. Dronov. 1984. Osobennosti struktury populyatsii soma Tsimlyanskogo vodokhranilishcha (Characteristics of the population structure of catfish of Tsimlyansk Reservoir). Ibid, No. 217, pp. 120–133.

Kuderskii, L.A. and Yu.I. Nikanorov. 1983. Vozrast kul'minatsii, ikhtiomassy i nastupleniya polovoi zrelosti v populyatsiyakh promyslovykh ryb Ivan'kovskogo i Uglichskogo vodokhranilishch (Age of ichthyomass culmination and onset of maturation in populations of commercial fish of Ivankov and Uglich reservoirs). Ibid, No. 202, pp. 133–152.

Kuderskii, L.A., A.S. Pechnikov and G.P. Rudenko. 1988a. Mnogoletnaya izmenchivost' vozrasta kul'minatsii ikhtiomassy v populyatsii ryb iz ozera pelyuga (Multiyear variability in the age of ichthyomass culmination in fish population from Pelyuga Lake). Ibid, No. 282.

Kuderskii, L.A. and O.I. Potapova. 1962. Gustera Lakshozera (White bream of Lakshozer). *Tr. Karel. Fil. AN SSSR*, No. 33, pp. 38–48.

Kuderskii, L.A. and G.P. Rudenko. 1982. Vozrast polovogo sozrevaniya i kul'minatsiya ikhtiomassy v populyatsiyakh massovykh vidov ryb malykh ozer Severozapada Evropeiskoi chasti SSSR (The age of maturation and ichthyomass culmination in population of numerous species of fish of small lakes of the Northwestern European USSR). *Sb. Nauch. Tr. Gos. NII Ozer. i Rech. Ryb Khoz-va*, No. 181, pp. 100–110.

Kuderskii, L.A. and G.P. Rudenko. 1988. Vozrast kul'minatsiya ikhtiomassy v populyatsiyakh ryb malykh ozer Severo-Zapada Evropeiskoi chasti SSSR (The age of ichthyomass culmination in fish population of small lakes of Northwestern European USSR). Ibid, No. 283, pp. 129–141.

Kuderskii, L.A., G.P. Rudenko and V.Ya. Nikandrov. 1983. Vozrast polovogo sozrevaniya i kul'minatsii ikhtiomassy v populyatsiyakh okun iz malykh ozer (The age of maturation and ichthyomass culmination in perch populations from small lakes). Ibid, No. 207, pp. 139–149.

Kuderskii, L.A. and L.M. Khuzeeva. 1986. Vozrast kul'minatsiya ikhtiomassy i massovogo polovogo sozrevaniya v populyatsiyakh khishchnykh ryb Kuibyshevskogo vodokhranilishcha (The age of ichthyomass culmination and mass maturation in populations of carnivorous fish of Kuibyshev Reservoir). Ibid, No. 250, pp. 12–29.

Kuderskii, L.A., L.M. Khuzeeva and K.S. Goncharenko. 1988b. Struktura populyatsii leshchi Kuibyshevskogo vodokhranilishcha (The population structure of bream of Kuibyshev Reservoir). Ibid, No. 280.

Kuderskii, L.A. and L.A. Yankovskaya. 1989. O polnote ispol'zovaniya rybnykh zapasov v vodokhranilishchakh Volzhsko-Kamskogo kaskada (On the total utilization of fish reserves in reservoirs of the Volga-Kama cascade). Ibid, No. 303.

Kulikov, E.F. 1966. Morfobiologicheskie osobennesti okunya malykh gumifitsirovannykh vodoemov Konchezerskoi sistemi (Morpho-biological characteristics of small humified water bodies of the Konchezer system). *Uch. Zap. Karel. Ped. In-ta*, Vol. 19, pp. 66–77.

Lapitskii, I.I. 1962. Uchet chislennosti e'ksplutiruemykh stad sazana, leshcha, i drugikh promysloovykh ryb Tsimlyanskogo vodokhranilishcha (Estimate of the population of exploitable stocks of carp, bream and other commercial fish of the Tsimlyansk Reservoir). In *Tr. Zonal'nogo Soveshch. po Tipologii i Biol. Obosnovaniyu Rabokhoz. Ispol'zovaniya Vnutrennykh (Presnevodnykh) Vodoemov Yuzhnoi Zony SSSR*, Kishinev, pp. 306–311.

Lapitskii, I.I. 1967. Metod ucheta chislennosti ryb v Tsimlyanskom vodokhranilishche (A method of estimating the number of fish in Tsimlyansk Reservoir). *Tr. Volgogr. Otd-niya Gos NII Ozer. i Rech. Ryb. Khoz-va*, Vol. 3, pp. 117–130.

Lapitskii, I.I. 1970. Napravlennoe formirovanie ikhtiofauny in upravlenie chislenost'yu populyatsii ryb v Tsimlyanskom vodokhranilishche (Planned formation of ichthyofauna and regulation of the population number of fish in Tsimlyansk Reservoir). Ibid, Vol. 4, pp. 1–280.

Lesnenko, V.K. and V.N. Abrosov. 1973. Ozera Pskovskoi oblasti (Lakes of the Pskov Region). Pskov, 154 p.

Mel'yantsev, V.G. 1949. Vliyanie distrofikatsii vodoemov na ikhtiofaunu (The effect of dystrophication of reservoirs on ichthyofauna). In *Prirodnye Resursy Istoriya i Kul'tura Karelo-Finskoi SSR*, Petrozavodsk, pp. 115–122.

Menshutkin, V.V. 1971. Matematicheskoe modelirovanie populyatsii i soobshchestv vodnykh zhivotnykh (Mathematical modeling of populations and communities of Aquatic Animals). Nauka, Leningrad, 196 p.

Menshutkin, V.V. and L.A. Zhakov. 1964. Opyt matematiches-kogo opredeleniya kharaktera dinamiki chislennosti okun v zadannykh e'kologicheskikh usloviyakh (Experience of mathematical determination of the nature of dynamics of perch population in the given ecological conditions). In *Ozera Karelskoago Peresheika*, Nauka, Moscow-Leningrad, pp. 140–155.

Monastyrskii, G.N. 1952. Dinamika chislennosti promyslovykh ryb (Population dynamics of commercial fish). *Tr. VNII Mor. Ryb. Khoz-va i Okeanografii*, Vol. 21, pp. 3–162.

Nebol'sina, T.K. 1960. E'kosistema Volgogradskogo vodokhranilishcha i puti sozdaniya ratsional'nogo rybnogo khozyaistva (Ecosystem of the Volgograd Reservoir and ways of estimating rational fisheries). Authors' Abstract of Dissertation for Doctor of Biological Sciences, Leningrad, 46 p.

Nebol'sina, T.K., N.S. Elizarova and L.P. Abramova. 1980. Vidovoi sostav ikhtiofauny, chislennost i zapasy ryb (Species composition of ichthyofauna, population and stocks of fish). In *Rybokhozyaistvennoe Osnovanie i Bioproduktsionnye Vozhmozhnosti Volgogradskogo Vodokhranilishcha*, Saratov, Izd-vo Sarat Un-ta, pp. 143–184.

Nebol'sina, T.K., L.P. Abramova, V.P. Ermolin and L.K. Smirnova. 1986. Metodika ucheta rybnykh zaposov i progmoza vylova v Volgogradskom vodokhranilishche (A method of estimating the fish stocks and forecast of catches in Volgograd Reservoir). *Sb. Nauch. Tr. Gos. NII Ozer. i Rech. Ryb. Khoz-va*, No. 244, pp. 70–79.

Nikanorov, Yu.I. 1984. Ivank'kovskoe i Uglichskoe vodokhranilishcha (Ivankov and Uglich reservoirs). Ibid, No. 210, pp. 4–12.

Nikol'skii, G.V. 1974. Teoriya dinamiki stada ryb kak biologicheskaya osnova ratsional'noi e'ksplutatsii i vosproizvodstva rybnykh resursov (Theory of Fish Population Dynamics as the Biological Background for Rational Exploitation and Management of Fishery Resources). Pishch. Prom-stv., Moscow, 2nd edition, 447 p.

Pechnikov, A.S. 1986. Dinamika chislennosti ryb i ratsional'noe ispolzovanie ikh zapasov v malykh ozerakh Severo-Zapada SSSR (Dynamics of fish population and rational utilization of their stocks in smaller lakes of Northwestern USSR). Author's Abstract of Dissertation for Candidate of Biological Sciences, Leningrad, 21 p.

Pechnikov, A.S. and I.I. Tereshenkov. 1984. K metodike sbora i obrabotki ikhtiologicheskogo materiala. Soobshch. I. Opredelenie razmerno-vozrastnoi struktury ulova (On the method of collection and processing of ichthyological material. Communication I. Determination of size- and age-structure of catch). *Sb. Nauch. Tr. Gos. NII Ozer. i Rech. Ryb. Khoz-va*, No. 225, pp. 90–100.

Pechnikov, A.S., I.I. Tereshenkov and A.E. Korolev. 1983. Opredelenie minimal'noi normy vylova malotsennykh ryb v zaryblennykh ozerakh (na primere ozera Naryadnogo) [Determination of the minimum size of catch of trash fish in overpopulated lakes (on the example of Naryadnoe Lake)]. Ibid, No. 198, pp. 205–220.

Perminov, L.G. 1981. Sostoyanie populyatsii leshcha nekotorykh ozer Pskovskoi oblasti v svyazi s zapuskom rybolovstva (The state of bream population of some lakes of Pskov Region in connection with launching fisheries). *Sb. Nauch. Tr. Gos. NII Ozer. i Rech. Ryb. Khoz-va*, No. 167, pp. 44–75.

Perminov, L.G. and Yu.V. Aleksandrov. 1986. Biologicheskoe obosnovanie regulirovaniya promysla leshcha na malykh ozerkh Pskovskoi oblasti (Biological basis of regulating bream fisheries in small lakes of Pskov Region). Ibid, No. 249, pp. 4–20.

Poddubnyi, A.G. 1971. E'kologicheskaya topografiya populyatsii ryb v vodokhranilishchakh (Ecological Topography of Fish Populations in Reservoirs). Nauka, Leningrad, 312 p.

Poddubnyi, A.G. and L.K. Malinin. 1988. Migratsiya ryb v vnutrennykh vodoemakh (Fish Migrations in Inland Reservoirs). Agropromizadat, Moscow, 225 p.

Poddubnyi, A.G., L.K. Malinin and V.G. Tereshchenko. 1982. O tochnosti otsenki absolyutnoi chislennosti ryb vo vnutrennykh vodoemakh (On the precision of estimates of the absolute numbers of fish in inland reservoirs). *Tr. In-ta Biologii Vnutrennykh Vod AN SSSR*, No. 49(52), pp. 82–102.

Popova, O.A. 1971. Biologicheskie pokazateli shchuki i okunya v vodoemakh s razlichnym gidrologicheskim rezhimom i kormnost'yu (Biological indicators of pike and perch in reservoirs with different hydrological regime and trophicity). In *Zakonomernosti Rosta i Sozrevaniya Ryb*, Nauka, Moscow, pp. 102–152.

Popova, O.A. 1979. Pitanie i pishchevye vzaimootnosheniya sudaka, okunya i ersha v vodoemakh raznykh shirot (Feeding and food interrelationships of pike-perch, perch and ruff in reservoirs of different latitudes). In *Izmenchivost' Ryb Presnevodnykh E'kosistem*, pp. 93–112.

Pronin, N.M. and A.N. Khokhlova. 1986. Kharakteristika parazitokhozyainoi sistemy trienoforus-okun v ozor Shu'chem (The nature of host-parasite system of *Trianophores*-perch in Shu'cho Lake). In *Issledovanie Vzaimosvyazi Kormovoi Bazy i Ryboproduktivnosti na Primere Ozer Zabaikal'ya*, Nauka, Leningrad, pp. 163–169.

Ressel, E'.S. 1947. Problema perelova ryby (The problem of Overfishing). Pishchepromizdat, Moscow, 94 p.

Riker, U.E. 1979. Metody otsenki i interpretatsii biologicheskikh pokazatelei populyatsii (A method of estimating and interpreting biological indices of population). Pishch. Prom-st', Moscow, 408 p.

Rudenko, G.P. 1962. Vozrastnyi sostav ryb, ikhtiomassa i ryboproduktivnost' okunevykh ozer (The age composition of fish, ichthyomass and fish productivity of perch lakes). *Nauch. Tekh. Byul. Gos. NII Ozer. i Rech. Ryb Khoz-va*, No. 16, pp. 33–37.

Rudenko, G.P. 1967. Opyt opredeleniya chislennosti ryb, ikhtiomassy i ryboproduktsii plotvichno-okunevogo ozera (Experience of determining the number of fish, ichthyomass and fish production in a roach-perch lake). *Izv. Gos. NII Ozer. i Rech. Ryb. Khoz-va*, Vol. 64, pp. 19–38.

Rudenko, G.P. 1971. Ikhtiomassa i chislennost' ryb v plotvichno-okunevom ozere (Ichthyomass and number of fish in a roach-perch lake). *Vopr. Ikhtiologii*, Vol. 11, No. 4(69), pp. 630–642.

Rudenko, G.P. 1978. Chislennost' ryb v malykh ozerakh Leningradskoi i smezhnykh oblastei i velichina ikh dopustimogo vylova (Number of fish in small lakes of Leningrad and adjoining regions and the magnitude of their permissible catches). *Izv. Gos. NII Ozer. i Rech. Ryb. Khoz-va*, Vol. 128, pp. 72–134.

Rudenko, G.P. and Yu.P. Volkov. 1974. Produktivnost' populyatsii ryb mezotrofnogo ozera Rachkova (bassein r. Velikoi) [Productivity of fish populations of the mesotrophic Rachkova Lake (Velikaya River basin)]. *Gidrobiol. Zhurn.*, Vol. 10, No. 4, pp. 84–89.

Rudenko, G.P., L.A. Kuderskii and A.S. Pechnikov. 1988. Potok e'nergii v ikhtiotsenozakh malykh ozer Severo-Zapada (Energy flux in ichthyocenoses of small lakes of the Northwest). *Sb. Nauch. Tr. Gos. NII Ozer. i Rech. Ryb. Khoz-va*, No. 282, pp. 126–135.

Sakharov, V.A. 1980. Vozrast i temp rosta sudaka ozera Il'men' (Age and growth rates of pike-perch in Lake Il'men'). *Ibid*, No. 159, pp. 94-97.

Sechin, Yu.T. 1977. Opredelenie chislennosti i ikhtiomassy leshcha ozera Il'men' (Determination of the population and ichthyomass of bream in Lake Il'men'). *Izv. Gos. NII Ozer. i Rech. Ryb. Khoz-va*, Vol. 126, pp. 83–92.

Sechin, Yu.T. 1979. Nekotorye rezul'taty rybokhozyaistvennykh issledovanii na ozere Il'men' (Some results of fisheries investigations in Lake Il'men'). *Ryb. Khoz-vo*, No. 11, pp. 37–40.

Sechin, Yu.T. 1986. Metodicheskie ukazaniya po otsenke chislennosti ryb v presenevodnykh vodoemakh (Methodological Notes on Estimation of Population of Fish in Freshwater Reservoirs). Vsesoyuz. Nauch.-Proizv. Ob-nie po Rybovodstvu, Moscow, 50 p.

Shibaev, S.V. 1986. Zakonomernosti funktsionirovaniya i puti ratsional'nogo ispol'zovaniya populyatsii leshcha Cheboksarskogo vodokhranilishcha (Regularities of functioning and ways of rational utilization of bream populations of Cheboksar Reservoir). Author's Abstract of Dissertation for Candiate of Biological Sciences, Leningrad, 21 p.

Shibaev, S.V. and V.I. Bandura. 1985. Razmerno-vozrastnaya struktura, smertnost' i produktivnost' populyatsii leshcha Cheboksarskogo vodokhranilishcha (Size and age structure, mortality and productivity of bream populations of Cheboksar Reservoir). *Sb. Nauch. Tr. Gos. NII Ozer. i Rech. Ryb. Khoz-va*, No. 240, pp. 53–63.

Tyunyakov, V.M., V.P. Koval' and L.f. Naumova. 1984. Dinamika chislennosti i intensivnosti promysla sudaka i bersha v Tsimlyanskom vodokhranilishche (Dynamics of number and intensity of fishing of pike-perch and Volga zander in Tsimlyansk Reservoir). Sb. Nauch. Tr. Gos. NII Ozer. i Rech. Ryb. Khoz-va, No. 218, pp. 53–59.

Tyurin, P.V. 1954. Biologicheskie obosnovaniya regulinovaniya sostava ryb vo vnutrennykh vodoemakh (Biological bases of regulating the composition of fish in inland reservoirs). In Materialy Soveshch. po Probl. Povysheniya Ryb. Produktivnosti Vnutrennykh Vodoemov Karelo-Finskoi SSR, Gosizdat Karelo-Fin. SSR, Petrozavodsk, pp. 62–75.

Tyurin, P.V. 1957. Biologicheskie osnovaniya rekonstruktsii rybnykh zapasov v Severo-Zapadnykh ozerakh SSSR (Biological basis of re-constituting Stocks in the Northwestern Lakes of the USSR). Izv. Gos. NII Ozer. i Rech. Ryb. Khoz-va, Vol. 40, pp. 1–202.

Tyurin, P.V. 1963. Biologicheskie osnovaniya regulirovaniya rybolovstva na vnutrennykh vodoemakh (Biological Basis of Regulating Fishing in Inland Reservoirs). Pishchepromizdat, Moscow, 119 p.

Tyurin, P.V. 1972. "Normal'nye" krivye perezhivaniya i tempov estestvennoi smertnostyu ryb kak teoreticheskaya osnova regulirovaniya rybolovstva (The "normal" survival curve based on rate of natural mortality in fish as a theoretical basis to regulate fishing). Izv. Gos. NII Ozer. i Rech. Ryb. Khoz-va, Vol. 71, pp. 71–128.

Tyurin, P.V. 1974. Teoreticheskie osnovaniya ratsional'-nogo regulirovaniya rybolovstva (Theoretical bases of rational regulation of fishing). Ibid, Vol. 86, pp. 7–25.

Umnov, A.A. and G.P. Rudenko. 1979. Opredelenie sredmei mnogoletnei chislennosti ryb, ikhtiomassy i ryboproduktsii v malykh ozerakh (biologo-matematicheskaya model' populyatsii) [Determination of mean multiyear population of fish, ichthyomass and fish production in small lakes (biomathematical population model)]. Gidrobiol. Zhurn., Vol. 15, No. 1, pp. 43–52.

Vetkasov, S.A. 1961. Promyslovo-biologicheskaya kharaktersika osnovnykh vidov ryb ozera Il'men' (Fishery-biological characteristics of the main species of fish in Lake Il'men'). In Zhivotnye Vodoemov Novogradskoi Oblasti, Leningrad, pp. 22–46.

Zaloznykh, D.V. 1985a. Struktura nerestovoi populyatsii, osobennosti rosta i plodovitosti shchuki Chelyabinskogo vodokhranilishcha (The structures of spawning population, peculiarities of growth and fecundity of pike of Chelyabinsk Reservoir). In Nazemnye i Vodnye Ekosistemy: Mezhvuz. Sb., Gork. Un-t., Gor'-kii, pp. 112–118.

Zaloznykh, D.V. 1985b. Struktura populyatsii i osobennosti rosta sudaka Chelyabinskogo vodokhranilishcha (Structure of population and characteristics of growth of pike-perch of Chelyabinsk Reservoir). *Sb. Nauch. Tr. Gos. NII Ozer. i Rech. Ryb. Khoz-va*, No. 240, pp. 64–72.

Zasosov, A.V. 1976. Dinamika chislennosti promyslovykh ryb (Dynamics of Numbers of Commercial Fish). Pisch. Prom-st, Moscow, 342 p.

Zonov, A.I. 1974. Instruktsiya po total'nomu oblovu malykh ozer (Instructions on total catches from small Lakes). *Gos. NII Ozer. i Rech. Ryb. Khoz-va*, Leningrad, 17 p.

Zhakov, L.A. 1962. O sposobe opredeleniya absolyutnoi chislennosti ryb metodom scheta okunevykh kladov (On a method of determination of the absolute number of fish by counting perch eggs). In *Gidrologicheskie Issledovania*, In-t Zoologii i Botaniki AN ESSR, Tartu, No. 3, pp. 353–357.

Zhakov, L.A. 1968. O prisposobitel'nom znachenii razmernoi i vozrastnoi struktury populyatsii okunya v malykh ozerakh Karel'skogo peresheika (On the adaptive significance of size and age structure of perch populations in the small lakes of the Karelian Isthmus). In *Syr'evye Resursy Vnutrennykh Vodoemov Severo-Zapada*, Karel. Kn. Izd-vo, Petrozavodsk, pp. 324–330.

Zhakov, L.A. 1974. Sostav i suktsessii ozernykh ikhtiotsenozov v svyazi s spetsifikoi faunisticheskikh kompleksov ryb (Composition and successions of lake ichthyocenoses in connection with the specifics of faunal complexes of fish). *Vopr. Ikhtiologii*, Vol. 14, No. 2(85), pp. 237–248.

Zhakov, L.A. 1984. Formirovanie i struktura rybnogo naseleniya ozer Severo-Zapada SSSR (Formation and structure of fish populations of the Lakes of Northwestern USSR). Nauka, Moscow, 144 p.

Printed in India.